WINGS ACROSS CANADA

WINGS ACROSS CANADA

An Illustrated History of Canadian Aviation

Peter Pigott

A HOUNSLOW BOOK
A MEMBER OF THE DUNDURN GROUP
TORONTO · OXFORD

Publisher: Anthony Hawke
Copy-Editor: Andrea Pruss
Design: Jennifer Scott
Printer: Friesens

National Library of Canada Cataloguing in Publication Data

National Library of Canada Cataloguing in Publication

Pigott, Peter
   Wings across Canada : an illustrated history of Canadian aviation / Peter Pigott.

Includes bibliographical references and index
ISBN 1-55002-412-4

1. Airplanes—Canada—History. 2. Aeronautics—Canada—History. I. Title.

TL523.P53 2002   629.133'34'0971   C2002-902287-8

1   2   3   4   5   06   05   04   03   02

We acknowledge the support of the **Canada Council for the Arts** and the **Ontario Arts Council** for our publishing program. We also acknowledge the financial support of the **Government of Canada** through the **Book Publishing Industry Development Program** and **The Association for the Export of Canadian Books**, and the **Government of Ontario** through the **Ontario Book Publishers Tax Credit** program.

Care has been taken to trace the ownership of copyright material used in this book. The author and the publisher welcome any information enabling them to rectify any references or credit in subsequent editions.

J. Kirk Howard, President

Printed and bound in Canada.✛
Printed on recycled paper.

www.dundurn.com

Dundurn Press
8 Market Street
Suite 200
Toronto, Ontario, Canada
M5E 1M6

Dundurn Press
73 Lime Walk
Headington, Oxford,
England
OX3 7AD

Dundurn Press
2250 Military Road
Tonawanda NY
U.S.A. 14150

This book is dedicated to Barbara and Guy,
the last of the New Delhi Pigotts

ACKNOWLEDGEMENTS

Everyone knows that Icarus came to a bad end attempting to fly, but how many people sympathize with his poor biographer trying to get eyewitness accounts and photos of the event? This is my ninth book on aviation, and I have learnt that much of what an aviation author does is not commune with the Muse but harass innocent people. In this case, I relied on the generosity of Captain Leah Gillespie, 4 Wing Public Affairs, Cold Lake Alberta; Major Lynne Chaloux, Senior Public Affairs Officer, 1 Canadian Air Division Headquarters, Winnipeg; and Janet Lacroix, CF Photo Unit, Ottawa, among others. On the civilian side, Ken Leigh came through once more, as did the National Archives in Ottawa and Peter Gauthier at Dollco Digital, who printed some of the photos. To all who helped, my gratitude.

TABLE OF CONTENTS

Avro 504K.

Courtesy of the Department of National Defence

AVRO 504K

The central landing skid, the wing tip hoops, and the hundreds of bracing wires made for an unattractive aircraft. The rotary engine's gyroscopic force would kill many student pilots. But the Avro 504, especially the K version, was built in great numbers, enough to continue on for a decade after the Treaty of Versailles was signed, to be used as the basic trainer of the RAF, and to become the preferred aircraft of barnstormers in the 1920s.

The Royal Flying Corps initially ordered the 504 as a training and reconnaissance aircraft, but in 1914, they pressed it into service as a bomber in the famous air raid on the airship sheds at Friedrichshafen. The idea of the 504 was to prepare the novice pilot to fly the more pernicious fighter aircraft of that day, like the Sopwith Camel and the Pup — both entirely unforgiving machines. The Avro 504's greatest advocate was Colonel Robert Smith-Barry, who had long complained that the high casualty rate among inexperienced pilots at the front was the result of the poor training they had received at home. As the new commanding officer at the School for Special Flying in Gosport, Hampshire in 1917, Smith-Barry ordered several Avro 504s, one of which was flown by the first Canadian air ace, Duncan Bell-Irving. In a letter to the family in Vancouver, Bell-Irving would write of the Avro: "I'm here instructing on what is quite the most perfect two seater aeroplane made. AVRO with its 100 hp Le Rhone rotary engine. They're quite out of date as regards to war but quite the most perfect thing to fly and teach people on."

There really wasn't much in the way of controls for the trainee pilot; the four basic instruments were the airspeed indicator, the altimeter, the inclinometer, and the engine speed indicator. On his left, the magneto switch and the fuel tank selection control stood out, as did the control column in the centre and the comparatively large compass that was positioned beneath the dashboard. Takeoff and landing were the most difficult to master, so the instructors kept these for the trainee to learn at the very last. He had no real control over it — it was either on or off. To land the aircraft, the petrol supply was turned off and then turned on again when the wheels were firmly on the ground, the fuel lever adjusted until the engine fired in short bursts for taxiing. If the mixture was too rich when the pilot was in the air, the engine lost power, and all the pilot could do was close the fuel lever off and look for a suitable field to land in. With its lightweight design, the Avro was sensitive to even a slight breeze, and the pilot had to land directly into the wind — probably the last thing on the poor pilot's mind at the time. The 504's tandem cockpit, with the instructor in the rear, where he could watch the poor student, was equipped with the latest means of communication, the Gosport tube, invented by Smith-Barry, which could be plugged into the pupil's helmet. Before its use, instructors communicated by passing notes, by yelling, or, as one hapless pilot recalled, by removing the control column and hitting him on the head with it.

Understandably, given the 504's propensity for spins, the instructors were afraid to give pupils charge of the aircraft. For the scourge of the 504's pilot was its engine, or various engines: the 110 hp Le Rhone, the 130 hp Clerget, and, most usually, the 110 hp Gnome Monosoupape rotary engine. Avro had designed the basic airframe to accept each. The rotary engine's gyroscopic force, combined with its weight in the nose, affected the 504's handling. It caused the nose to rise on left turns and drop with right, which required compensation by the rudder. Such was the gyroscopic effect that left or right turns needed full left rudder, and the poor pilot was always battling the force from the engine's torque, which took the aircraft sharply to the right. Little was known about spins then, and many a novice pilot was killed by his inability to recover from an unintentional one. But at a time when many aircraft had rotary engines, this was a common ailment, and the Avro 504 handled better than most — notably the famous Sopwith Camel, which shared the same 110 hp Gnome and 110 hp Le Rhone.

ARMSTRONG-WHITWORTH SISKIN

In 1922, the Royal Air Force looked to replace its Sopwith Snipes, which dated from the Great War. The air force requirement called for a fighter with an all-metal airframe and a fabric covering, and the Armstrong-Whitworth (AW) Company bid on the contract. It had had some experience during the First World War with its F.K. 8 and F.K. 10 armed reconnaissance biplanes, which were sturdy, if gawky, aircraft. Now, looking to re-establish itself in the aviation world, AW bought the Siddeley Deasy aero engine company and then designed another fighter aircraft.

The company's first effort was a single-seat wooden sesquiplane with an Armstrong Siddeley Jaguar radial engine. Named the Siskin, it competed for the RAF order with the Gloster Grebe fighter. The RAF bought sixty-four Siskin IIIs in 1924 and four hundred of the model IIIAs and Bs three years later. Manoeuvrable and spirited, the Siskins were popular with the air force, and some were exported to Estonia, where they were still in use when the Second World War began. With this investment, the company expanded into the commercial field with its three-engined Argosy for Imperial Airways.

Armstrong-Whitworth had sent a pair of Siskins to Canada for winter testing, and in 1927, the Canadian government allowed the RCAF to purchase nine Siskins and six Atlases. These were the first new fighter aircraft since the Great War and, at the time of purchase, the latest in technology. The Atlases were for an army cooperation role, but the Siskins were sent to No. 1 Fighter Squadron. When assembled at Camp Borden, three of them were designated as part of the RCAF's first aerobatic team: the Siskin Flight, which performed at air shows until the mid-1930s.

The RCAF was starved of aircraft purchases after that, and the Siskin Flight was the closest that the poor RCAF pilots got to combat conditions. The Siskins laboured through the decade with No. 1 Fighter Squadron at Trenton until the purchase of the Hurricanes in 1939.

Armstrong-Whitworth Siskin at Rockliffe.

Courtesy of the Department of National Defence

WESTLAND WAPITI

The early 1930s were characterized by two trends worldwide: the Depression and disarmament. The first meant that there was little money to buy new aircraft for the RCAF, and the second gave a reason not spend what money there was. As a result, even with the rise of Fascism in Europe, in 1935 the Canadian government could afford only second-hand, inferior fighter aircraft like the Westland Wapiti.

The return to power of Prime Minister Mackenzie King in 1935 allowed for the reorganization of the air force to provide for some token coastal defence. The RCAF, like the two other services, was tied to its British counterpart (the RAF) for organization and equipment. Accordingly, although the purchase of American aircraft would have been cheaper, Canada continued to look toward Britain for its equipment. In 1935, Ottawa ordered six Westland Wapiti bombers and four Blackburn Shark torpedo bombers. Neither aircraft was adequate for Canadian purposes; both were second-hand from the RAF. The Westland Wapiti biplane had been named after the North American elk — appropriately, for like that creature, it was defenceless. The aircraft itself was little more than a copy of the De Havilland 9A of Great War vintage. It was underpowered, poorly designed, "glided like a brick," and had little to recommend itself. The RAF had found it wanting even for use at its stations in the Middle East and was pleased to unload it on the penniless Canadians. The biplanes, having braved the desert and tribesmen's bullets, arrived in Canada in deplorable condition. The contract to clean their cockpits of scorpions and camel dung (if the hapless RCAF pilots were to be believed) was given to an Ottawa street car company that was owned by the Ahearn family, which, by coincidence, had been generous supporters of the Liberal party's re-election.

Westland Wapiti.

National Archives of Canada PA 63307

But the Wapitis (nicknamed "What a pities") were Canada's only bomber in the 1930s and were assigned to RCAF No. 3 Squadron. Based at Ottawa, they took part in military exercises at Camp Borden until 1938, when No. 3 was moved to Calgary to better train on bombing runs. With the war, the RCAF gratefully demoted the Wapitis to training and observation roles.

National Archives of Canada PA 63J19

Wapitis at Ottawa, the RCAF's only bomber in the 1930s.

Tiger Moth.

THE TIGER MOTH

Given birth by the Golden Age of flying in the 1920s, the De Havilland Tiger Moth became the basic pilot trainer for Commonwealth air forces during the Second World War, serving with the Royal Air Force until 1953. When, between the world wars, flying developed into a fad for youthful aviation enthusiasts (both male and female) the director of British Civil Aviation, Sir Sefton Branckner, looked for a cheap, uncomplicated airplane with which to launch a national flying club scheme. And so was born one of the most beloved biplanes in aviation history: the De Havilland Tiger Moth.

Geoffrey de Havilland said that he built the two-seat biplane for his own use and, as a natural history hobbyist, thought the name "Moth" perfect, keeping to that family of insects to name the whole series of light aircraft. On February 22, 1925, he test flew the first Moth, the DH 60 Cirrus, himself. It wasn't only the aircraft's robust simplicity that attracted the public, but the fact that it was an affordable £885, the price of a car — and indeed, with the wings folded, it could be towed by the family car. It was the first "people's plane," and it was said anyone could learn to fly or service a Moth, so uncomplicated were its controls and engine. Branckner agreed and ordered ninety, putting De Havilland into the main ranks of British aircraft builders. The Royal Air Force had misgivings about its lack of forward visibility and the maze of struts that hampered access to the front (or student's) cockpit. To overcome both these defects, the DH 60 evolved into the DH 82 Tiger Moth, which would fly on October 26, 1931. Distinguished by its swept-back wings for improved cockpit access and powered by a 120 hp inline Gipsy four-cylinder engine (which by 1937 had been upgraded to a 130 hp engine), it cruised comfortably at 100 miles per hour, had a ceiling of 13,600 feet, and had a range of 285 miles. In fine weather, the Tiger Moth was a delight to fly. By the time the war began, 1,153 Tiger Moths had been built, and it was the Royal Air Force's basic trainer.

Geoffrey de Havilland would never have guessed how many millions of pilots worldwide would learn to fly on his aircraft before moving on to Harvards and Ansons, and ultimately Spitfires, Lancasters, and commercial airliners. Simple enough to be built by even the most basic aircraft industry, Tiger Moths were made in four dominions (with a few variations), as part of the British Commonwealth Air Training Plan. De Havilland Canada turned out 1,747 Tiger Moths as the DH 82C, distinguished by their enclosed canopies; some had 160 hp Pirate D.4 engines. There were 345 manufactured in New Zealand, and in Australia, the 1,085 Tiger Moths built locally were known as Jackaroos. Of the many historic flights that Tiger Moths made in Canada, one of the most colourful took place on June 24, 1968. To commemorate the fiftieth anniversary of Canada's first airmail flight, Trans Canada Airlines captain Donald Chamberlin and ex-RCAF pilot Thomas Lee re-enacted the famous flight with a Tiger Moth. The original pilot, Captain Brian Peck of the RAF, had used a Curtiss Jenny to fly 120 letters in a mail sack from Montreal's Cartierville Airport (then Bois Franc polo grounds) to Toronto's Leaside airfield, completing the trip in seven hours. A co-owner of the Tiger Moth, Chamberlin took same number of letters in the original mailbag from thirteen-year-old Brian Peck, Captain Peck's grandson. The Tiger Moth made the flight in five hours, with the same stops at Kingston and Desoronto, and landed at Malton Airport, as Leaside had become a suburb of Toronto.

FOKKER UNIVERSAL

The first bushplanes in Canada were war surplus HS-2L biplane flying boats, which were restricted to use in the short summer months. For year-round operation, what was needed were radial-engined, wooden-winged monoplanes that could be adapted to the season and used with wheels, floats, or skis. Anthony Fokker's Universal was just that.

The Dutchman's business was flourishing during the 1920s, especially in the United States. Because of that country's late entry into the Great War, American aircraft designers had little experience in aircraft. Fokker did, and his planes were constantly used in record-breaking flights — the first across the United States non-stop in 1923, the first to fly from Amsterdam to Batavia (now Indonesia) in 1924, the first over the North Pole with Admiral Byrd in 1926, and the first flight from the mainland United States to Hawaii in 1927.

Initially importing his aircraft into the United States, in 1926, he opened Fokker Aircraft Corp in Teterborough, New Jersey to build exclusively for the North American market. The Fokker Universal was his masterpiece and set a standard in aviation in the 1920s. At a time when commercial aircraft were fragile fabric and wood biplanes with water-cooled engines, this highwing monoplane sported a very thick, strong, one piece plywood wing and a welded steel fuselage. Its Wright J-4B 200 hp radial engine gave it a maximum speed of 118 miles per hour and a range of 535 miles. Best of all for the Canadian bush, the Universal had an interchangeable undercarriage for skis, floats, or wheels. It was also able to land on rough ice or ground because of an elementary shock absorber — a coil of bungee cord. The pilot still sat in the open to brave the elements, but his four passengers were in an enclosed cabin.

Fokker Universal on floats.

To begin Western Canada Airways (WCA) in 1926, Winnipeg grain merchant James Richardson bought three Universals, investing the huge sum of one hundred thousand dollars in them. Flown up from New York by the company's first pilot, Harold "Doc" Oaks, they were christened "City of Winnipeg" (G-CAFU), "City of Toronto" (G-CAGD), and "Fort Churchill" (G-CAGE). The Universals initiated year-round passenger and freight service in Canada. The only innovation the Fokkers needed were new skis; the American skis that the aircraft came with were found unsuitable for local conditions and were replaced with Canadian skis made of ash, shod with brass, and weighted. The ski design would be standard on all WCA aircraft and was later used by Admiral Byrd on his Antarctic flight.

The company's first major contract came in January 1927, when the federal government asked if it could fly supplies and men to Fort Churchill on Hudson Bay. A government scheme to ship grain and cattle to Europe through Hudson Bay in the summer months, it depended on a railway from the prairies. With the help of British engineers, by 1926 the railhead had reached the swamps around Cache Lake, 350 miles from the port of Churchill. Exploratory drilling to deepen the harbour for ocean vessels could only be carried out while it was still frozen, hence the urgency for the supplies. Richardson hired three experienced pilots, J.R. Ross, Fred J. Stevenson, and the great Norwegian pilot Bernt Balchen, to fly his new Universals. The initial air shipment was thirty tons and fourteen passengers. Some of what the pilots were to carry was dynamite, considered a problem then because no one knew what effect air pressure would have on the explo-

Fokker Universal on skis.

Courtesy of the Schade family

sive. But from March 22 to April 17, 1927, using a boxcar for sleeping accommodations at the railhead at Cache Lake, WCA's pilots made twenty-seven round trips in the three Fokker Universals, each time landing on the frozen Churchill waterfront. It was the first airlift in history, and it gained both Richardson and Fokker fame and more business.

With the port of Churchill viable, the federal government then set about exploring and mapping the Hudson Straits for navigation by ocean-going ships. That May, the public's attention was

Fokker Universal with broken ski, from various angles.

Cause of the accident.

focused on Charles Lindbergh's Atlantic crossing, and the Hudson Straits expedition was overshadowed, but it ranks as one of early aviation's greatest enterprises. The party set out from Halifax on July 17, 1927, in two ships that carried forty-four men and enough supplies and fuel for a year and a half. On board, in specially built crates, were six Fokker Universals that the Department of Marine and Fisheries had bought. Flown by RCAF pilots, through the winter and the summer of 1927 and 1928, the Universals operated from base camps at Ungava Bay, Wakeman Bay, and Nottingham Island, radioing information on the terrain, ice, and weather to assist in the mapping. It was the first time that aerial operations on such a large scale had been conducted so far north. Despite the primitive conditions and hard use, all the Fokkers returned with the expedition to Ottawa in November 1928, all in working condition, their efforts recorded on film that would be discovered three decades later.

Richardson would buy a dozen Fokker Universals in total, making them the first workhorses of his fleet. Five would be involved in crashes — tragically, the first "Fort Churchill" killed Stevenson, after whom Winnipeg's airport was named. Now Canadian Airways Ltd., the company sold off the Universals to small operators like Arrow Airways. On November 30, 1931, G-CAGD "City of Toronto," now painted blue, was bought by Grant McConachie, the future bête noire of Canadian Pacific Airlines, for his one aircraft company, Independent Airways.

Unfortunately for Anthony Fokker, the popularity of his aircraft in the United States ended abruptly. On March 31, 1931, the wooden wing of a TWA Fokker F10 en route from Kansas City to Los Angeles fell off in flight. The aircraft crashed, killing all on board, one of whom was the revered coach of the "Fighting Irish" football team, Knute Rockne. In the media frenzy that followed, Washington ordered that all passenger-carrying aircraft be twin-engined and made of metal. This gave American aircraft manufacturers like Donald Douglas and Bill Boeing the impetus to build such planes. Fokker's company was bought by North American Aviation in 1933.

Fairchild FC-2W-2 "G-CART."

Courtesy of the Woollett family

FAIRCHILD FC-2 AND W-2

American inventor Sherman Fairchild came to aviation through aerial surveying. In 1922, his Fairchild Aerial Survey company was contracted by the Laurentide paper company for forest surveying at Grand Mere, Quebec, using a Curtiss Seagull flying boat. Fairchild knew at first hand the conditions that the pilots faced and found the Seagull (and later the HS-2L flying boats) inferior for surveying. The cameras were unwieldy, their users needed protection from the cold and yet as much visibility as possible around them, and the aircraft were incapable of year-round service and restricted to use on the water. Looking for a stable platform for aerial surveying, Fairchild designed his own aircraft, calling it the "All Purpose Monoplane."

Built at the Fairchild Airplane Manufacturing Co. in Farmingdale, New York, the FC-1 (for Fairchild Cabin) was everything that an aerial photographer needed. Powered by a 220 hp Wright J-5 Whirlwind engine, its body was made of welded steel tubing that narrowed to three longerons, earning the aircraft the name "razor back." Its undercarriage was interchangeable for wheels, skis, or floats, and it was the first aircraft to have Bendix hydraulic brakes. For aerial photography, Fairchild put in as many windows as he could, and, in a first for commercial aviation, the large windshield was shatterproof. But what really distinguished the FC-1 from the rival Fokker Universal was the heated, enclosed cabin for the pilot and four passengers. Another unique feature was the folding wings: two men could fold the forty-four-foot wingspan into a thirteen-foot unit for easier (and cheaper) storage. Capitalizing on the success of the FC-2, Fairchild then built the larger FC-2W-2, which could accommodate seven passengers and had a more powerful Pratt & Whitney Wasp engine. Because it had four longerons, it was called a "turtle back."

RCAF Fairchild FC-2W-2.

Author's Collection

So durable were Fairchild's planes that Admiral Byrd took an FC-2, christened the "Stars and Stripes," to Antarctica with him in 1927 and left it there. Five years later, the second expedition recovered it out of the ice and flew it, taking it back home in 1934, where, after much use and renovation, it is now on display at the Smithsonian National Air & Space Museum. Pan American Airway's first airmail flights were by FC-2 in 1928 between New York and Miami. The first regular United States–Canada cross-border service was conducted by Canadian Colonial Airways using FC-2W-2s between Montreal (St. Hubert) and New York. The RCAF bought Fairchilds of both models for communication and survey work, modifying them with Armstrong Siddeley Lynx engines. Canadian Airway's James Richardson was a good customer, buying the first of seven FC-2s in 1927 and eight FC-2W-2s through the 1930s.

Many of the bush pilots flew Fairchilds, especially the storied "Flying Postmen" of Quebec, who took off from the Quebec City suburb of Ste. Foy to conduct a winter airmail service to isolated communities. None were as flamboyant as Walter (Babe) Woollett. A former Royal Air Force pilot, Woollett immigrated to Canada in 1928 to work for Fairchild Aerial Surveys at Lac a la Tortue. A strong candidate for Canada's Aviation Hall of Fame, Woollett, in retirement in Hawaii, penned a witty autobiography, *Have A Banana*, recounting his exploits in the FC-2W-2 (G-CART) and later the Fairchild 71 (CF-AAX).

Courtesy of U.N.N.

Fairchild FC-2W-2 with pilot and engineer; note stepladder and funnel for pouring in heated oil to get the engine started.

Courtesy of the Woollett family

Fairchild in the air.

Courtesy of the Woollett family

Fairchild at Fort George, Quebec.

Fokker Super Universal.

FOKKER SUPER UNIVERSAL

In 1927, its first year of business, Western Canada Airways carried a staggering 420,730 pounds of freight and 1,141 passengers. The figures proved James Richardson's belief that aircraft could open up the north for mineral exploration. His company needed not just more aircraft but machines specifically designed for the Canadian climate and bush conditions, with more reliable, air-cooled engines, metal fuselages, and greater cargo-carrying capacity. The three Fokker Universals used in the Churchill airlift were soon too small and underpowered, and Richardson purchased fourteen Super Universals. The competition between Fokker and Fairchild in the 1930s was as intense as that between Boeing and Douglas in the 1950s, each superseding the other with a new model. The aircraft was built under licence by Canadian Vickers in Montreal, its Pratt & Whitney Wasp engine of 420 hp giving it double the power and endurance of its predecessor. It had a range of 675 miles and a higher ceiling than the Universal (18,000 feet compared with 11,500 feet). Most importantly, the pilot and six passengers were now enclosed from the elements. The first of the Fokker Super Universals arrived at the WCA base in Winnipeg in the summer of 1928. With these, WCA pilots could pioneer flights deep into the far north. On August 12, 1928, A.H. Farrington flew his Super Universal more than one thousand miles along the west coast of Hudson Bay from Winnipeg to Eskimo Point to pick up a prospecting party. Without aircraft, such a trip over land would have taken three months.

In 1929, a prospector called Gilbert Labine convinced Richardson that there were large mineral deposits in the Great Bear Lake region of the Northwest Territories. Wilfred Leigh Brintnell was flying Super Universal G-CASK to the Yukon and took Labine to the lake. When C.H. "Punch" Dickins picked him up three weeks later, Labine had discovered the richest deposit of pitchblende in history, and his Great Bear mine was responsible for many years of silver, radium, and uranium.

The MacAlpine Expedition and its rescue were feats of endurance that tested both men and aircraft. The president of Dominion Explorers, Lieutenant Colonel C.D.H. MacAlpine, planned to explore the Canadian Arctic in 1929 to assess the viability of mineral exploration. The year before, Dickins had flown him in G-CASK on a similar trip, covering four thousand miles in twelve days. This time, to transport a party of seven mining engineers, MacAlpine used his own Fairchild FC-2W-2, CF-AAO, and leased the WCA Super Universal G-CASP with G.A. "Tommy" Thompson as pilot and A.D. Goodwin as engineer. Aware that the trip had to be accomplished before winter set in, the two aircraft left Winnipeg for Hudson Bay at 10:00 A.M. on August 24, 1929, flying north to Norway House. With fuel low, they landed offshore at Churchill harbour on the twenty-sixth to await the supply ship *Morso*. In what would be the first in a series of calamities, they discovered that the schooner had caught fire at sea and blown up because of a cargo of dynamite. That night, G-CASP dragged its anchor and was swept into the sea by the tide. Its float torn open by ice, the aircraft sank, beyond repair. They radioed their plight to the WCA office in Winnipeg, and Francis Roy Brown flew G-CASK out on September 6. Both machines were loaded up, and the party continued on to a fuel cache at Baker Lake, where condensation in G-CASK's fuel tank caused frost particles in its radiator, and the party was delayed. They flew on and passed Pelly Lake, where they saw that the lakes that were to be the landing areas had started to freeze over earlier than expected. Faced with being trapped overnight in the ice, the pair of aircraft made for the open sea of the Arctic coast, hoping to land at one of the coastal settlements, where they could get a ship.

There was less than a few hours of fuel in either of the aircraft when the coast appeared. Both aircraft made it to Dease Point on the mouth of the Koolgaryuk River and set down there, marooned until help arrived. Everyone was reassured that it was still early September, with a month before the freezing temperatures were to begin. The crew built a small house out of stones, moss, and parts of the Fairchild's wing, with an engine cowling serving as a stove. On September 12, they tried to get G-CASK into the air with the few pints of fuel left, but once more its carburetor gave trouble. The nearest settlement was a Hudson's Bay post at Cambridge Bay, estimated by visiting Natives to

Courtesy of the Schade family

"G-CASK" with pilot and passenger.

Photo taken in 1932 of children at Eskimo Point, District of Keewatin, by pilot of "G-CASK."

be three days' walking distance — but only if the shoreline ice was strong enough to bear their weight. By October, when the first of the snowstorms struck on schedule, there were still no rescue aircraft to be seen, and food rations were almost gone. The Native diet of dried fish, while generously shared, gave the white men severe cramps. It was not until October 18 that the ice was judged thick enough by the Natives to cross along the coast, and two days later, employing Native guides, the party began its trek to Cambridge Bay. They made it to the Kent peninsula on October 24; across the Queen Maud Gulf was the outpost of Cambridge Bay. By now they had run out of food and sent the Natives back to Dease Point for more. They returned with the encouraging news that while they were there, an aircraft had flown over the hut.

Both WCA and Dominion Explorers mounted aircraft searches for the party, despite the imminence of the fall freeze-up period, when no flying was usually possible. Two search teams were organized: a main one with five floatplanes to operate as far north as they could, and a secondary group that would ferry fuel and supplies to them. Brintnell was put in charge of the overall search, and he, C.H. Dickins, Francis Roy Brown, Andy Cruickshank, Bill Spence, and Herbert Hollick-Kenyon searched the vast Mackenzie district, flying as far west as Fort Norman and Coppermine, and as far east as Baker Lake.

It was a race against the winter, as more of the inland water froze below the searchers daily, the crews very careful not to get trapped in it wherever they set down overnight. By October 13, all the lakes were frozen and the conversion to skis began. Baker Lake was used as a base, with all five aircraft hauled up on shore and fitted with skis. A severe gale lashed the thin ice against the beach on the seventeenth, damaging the aircraft, and it would not be until October 24 that four were serviceable enough for the search to continue. This time they reached the coast and landed their aircraft at a fuel cache on the frozen Burnside River. Three made it down safely, but Andy Cruickshank in Fokker G-CASQ misjudged the strength of the ice, and it plunged through, settling nose down in the water with only the wings showing. It was abandoned until the MacAlpine party could be found, and the other aircraft pressed to the Arctic Sea, flying over the now-empty hut at Dease Point.

At Burnside, despite sub-zero conditions and a cruel blizzard, G-CASQ was raised by block and tackle, the engine overhauled, and the aircraft made ready to resume the search when, on November 5, a Native dog team arrived with the news that the MacAlpine party had been found. The schooner *Fort James* at the Magnetic North Pole had intercepted a message from Amundsen's old ship, the *Bay Maud*, which was being used as a Hudson's Bay Company supply depot at Cambridge Bay. On November 3, at 4:30 P.M., with the temperature at twenty-seven degrees below zero, the MacAlpine party had staggered into the post, starving and suffering from frostbite. The four rescue aircraft landed alongside the *Bay Maud*, and the men were flown out in relays. But with the ferocity of winter, it would not be until December 4 that the entire MacAlpine Expedition party could be landed at the railway station at Cranberry Portage for the train to Winnipeg. Although the sunken G-CASP was insured and G-CASK recovered and made airworthy, WCA made little money on the epic flights. But much of northern Canada had been flown over for the first time, and the heroism of WCA's pilots and the durability of their Super Universals is worthy of recognition.

If this wasn't enough, the next year the Super Universal G-CASK would be involved in another flight into the high Arctic. Major L.T. Burwash was leading a government expedition to record the magnetic properties near King William Island and to check on the location of the Magnetic Pole. He was also anxious to find out what happened to the ill-fated Franklin expedition, which had disappeared in the region eighty years before. Pilot Walter Gilbert and engineer Stan Knight flew Burwash and his party in the Fokker up the Arctic coastline, overflying the Magnetic Pole. Gilbert was later honoured with a Fellowship in the Royal Geographic Society and the Trans Canada (McKee) Trophy.

Brintnell would leave Richardson in 1932 to begin his own air company, Mackenzie Air Services, also using Super Universals. Sadly, the following year, Canadian Airway's most famous Super Universal, G-CASK, would be accidentally set on fire at Fort McMurray. A total of eighty Super Universals were built, twenty-nine by Nakajima in Japan.

Fokker Super Universal.

JUNKERS W33/34

T he story of the Junkers W33/34 is the story of bush flying in Canada in the late 1930s. While there already were Fokkers and Fairchilds serving the many bush companies when the first Junkers came to this country in 1929, the German aircraft set a new standard in commercial aviation. It will forever be connected with the great bush pilots: "Con" Farrell, "Wop" May, Matt Berry, and Vic Horner.

Dr. Hugo Junkers was a German metallurgist who worked with Anthony Fokker during the First World War, building aircraft. On June 26, 1919, less than a year after the war's end, Junkers had designed, built, and flown the F13 monoplane from his factory at Dessau. Sometimes called the "Mother of all Passenger Planes," the F13 was the world's first all-metal airliner. Rather than being exposed to the elements in a wood and fabric biplane, now four passengers sat in a heated metal cabin made of his specially invented corrugated "Elektron" alloy. In 1926, Junkers entered the cargo market with the larger W33 and 34, flying them both within a month of each other. With bigger cabins than the F13 and hatches instead of windows, these were the original "flying boxcars"; the only difference between the two was that the W33 was powered by the inline, liquid-cooled Junkers L5 engine, while the W34 was offered with whatever engine the customer wanted (initially the Gnome and Rhone Jupiter VI). German pilots broke long distance records with both, and one, "Bremen," made the first successful east-west Atlantic crossing on April 12, 1928, landing at Greely Island, Labrador. Sold to customers as far away as Iceland, Iran, New Guinea, and Peru, a number of W33s were bought and made in the Soviet Union and Sweden to circumvent the Treaty of Versailles ban on the building of aircraft in Germany that had a military application. Impressed by the "Bremen," Canadian Airways owner James Richardson bought eight W33/34s (AQW, ASI,

The Junkers CF-AMZ, on the beach at Churchill in 1932, was a sister ship of CF-ATF, which TCA staff Greg Hoban and John Hutchison made airworthy and F/O John Pacey flew to Ottawa.

Art Schade photo

Above: Junkers W34 on skis.

Below: Same Junkers W34 on wheels.

ABK, AMZ, AQV, ARI, ASN, and ATF), one of which was made in Sweden. They arrived just in time for his company, as not only were the Fokker F14s too fragile for the bush, but several of his aircraft had been destroyed in a hangar fire at Stevenson Field, Winnipeg on March 4, 1931. Until that point, the cost of air freight was prohibitive for the people who needed it the most: the ordinary prospector and trapper. But with the Junkers, Canadian Airways could carry anything — mining equipment, mail bags, fuel drums, canoes, sled dogs — comparatively cheaply. In fact, knowing that they could only afford a single flight to stock up on winter provisions, the customers tended to overload the Junkers. Its all-metal body endured punishments that fabric could not have. Canoes could be carried beneath its wings, one on either side.

Whether from hard usage or the unforgiving elements, three Junkers crashed: CF-ASI at Kamainagami Lake, Ontario; CF-AQV at Gold Pines, Ontario; and CF-ARI at Waterways, Northwest Territories. One was retired in 1940 and reduced to spares to keep the others working. But what killed the aircraft was not the climate, the terrains, or the pilots. The federal government first cancelled the airmail contracts, forcing Canadian Airways to the edge of bankruptcy, and then began its own airline, Trans Canada Airlines, to carry the mail. The death of Dr. Hugo Junkers in 1935 and the Nazis coming to power in Germany allowed the company to build bomber versions of the W34 and use them in Spain. Richardson stopped buying Junkers aircraft then and turned instead to purchasing Canadian-built Noorduyn Norsemen in 1936.

The last surviving Junkers W34, CF-ATF, was still flying in 1962, thirty years after it had arrived in Canada. James Richardson's widow bought it and donated it to the National Aviation Museum in Ottawa that year, and it was flown

Courtesy of the Schade family

Earning its name: "Flying Boxcar."

across the country by an Air Canada crew. On the way to Ottawa, it stopped at Kamloops, Edmonton (Cooking Lake), Winnipeg, Fort William, Sault Ste. Marie, and Pembroke, everywhere evoking the glory days of bush aviation. On September 17, 1962, Walter Dinsdale, the national resources minister, greeted CF-ATF as the old Junkers came in for her final landing at Rockliffe, where she can be viewed today.

Courtesy of the Schade family

Demonstrating the strength of the wings on the Junkers.

FAIRCHILD 71

The FC-2W-2 was so effective, especially in the bush, that Sherman Fairchild not only refined it as the Model 71 but also had it built in Canada for local conditions. He established Fairchild Aircraft Ltd. in the village of Longueuil, eight miles from Montreal, to do so. Here, a total of twenty-one Model 71s in "B" and "C" versions were built, each a variation of the FC-2W-2 in size and strength. The 71C would be the first Fairchild to have a metal-covered fuselage.

The RCAF bought twelve Fairchild 71Bs for photography and transport, using them from 1930 to 1941 then selling them to Canadian Airways, which, because of the war, was desperately short of aircraft. Richardson had some of his FC-2W-2s converted to 71s; G-CAVV, which crashed in 1938, was rebuilt as a 71C and re-registered as CF-BJE. Flown by pilot Art Schade on "treaty flights," it operated out of the Canadian Airways base at Sioux Lookout, Ontario. These were federally sponsored flights; under the terms of the native treaties, a party made up of an RCMP constable, a doctor, a justice of the peace, and an official from the Department of Indian Affairs met annually with aboriginal tribes to dispense the law, medicine, and money.

Fairchild continued the family with the Super 71 in 1935 and the Model 82 the following year. By the Second World War, most of the 71s had been reduced to spares or had crashed; the few remaining were used by the Ontario Provincial Air Service until the advent of the De Havilland Beaver. Sherman Fairchild died in 1971. Rather appropriately, one of his aircraft would figure in the Canadian bush again. In the 1950s, his C-119 Flying Boxcars would be used by the RCAF to build and supply the Mid-Canada Radar Line.

Courtesy of the Schade family

Fairchild 71.

Courtesy of the Schade family

Fairchild 71 on treaty flight.

Courtesy of the Schade family

July 1937 treaty flight, with tribe gathered to meet plane.

Courtesy of the Schade family

RCMP constable, doctor, Indian agent, and pilot on treaty flight.

Junkers 52/1m "CF-ARM."

JUNKERS 52/1M

Goering had three, all painted red, all named after his idol, Manfred von Richthofen. Hitler was presented with one in 1934 and cherished it. When his staff tried to upgrade him to the four-engined Focke Wulf Condor, he showed no interest and then angrily refused, saying he was used to his dependable old Ju 52. The aircraft that would bomb Guernica, drop paratroops over Crete, and reinforce the German Army at Stalingrad survived the war to be built in France and flown by British European Airways on its London to Belfast route. Gawky, slow, noisy, and draughty, the Ju 52 was known to its German crews as "Tante Ju," "Iron Annie," or, in the later part of the war, "The Corrugated Coffin." Even now, wrecks are found in the jungles of South America and New Guinea, and in Norwegian fjords.

Dr. Hugo Junkers held that there was a sufficient market for such a large cargo plane and presented his corrugated metal, cantilever monoplane Ju 52/1m (for single engine) to the public at Tempelhof Airport on February 3, 1931. But both Lufthansa and the Luftwaffe wanted an aircraft of that size to carry seventeen passengers, troops, or bombs rather than freight, and persuaded Junkers to convert it to the three-engined Ju 52/3m, the first of which flew on March 7, 1932. Thus, only seven of the single-engined version were built, and only one was exported — to Canada.

When Prime Minister R.B. Bennett canceled the federal mail contracts in 1930, James Richardson looked to increase his company's cargo-carrying capacity instead. Impressed by the versatility of his Junkers W33/34s, he purchased Junkers's giant single-engined Ju 52/1m. With its corrugated duralumin skin (unable to obtain aluminum, Junkers developed his own "Elektron" metal) and three main wing spars, the Ju 52 was ideal for bush conditions. Its large flaps allowed it to

Both Junkers at Collins, Ontario, 1934.

land in restricted areas. Its massive undercarriage tolerated takeoffs from the most primitive airfields. It could carry its own weight into the air (from three to eight tons), and its systems were so elementary that it could be repaired in the bush — the only hydraulics were the brakes.

The choice of power plant was optional, and Richardson sent his engineering manager, Tommy Siers, to tour the Junkers plant at Dessau, Germany. Siers decided to replace the standard Junkers L88 engine with a 12-cylinder, liquid-cooled, 685 horsepower Bavarian Motor Works (BMW) engine. The Junkers was shipped to Montreal on the S.S. *Beaverbrae* and, when assembled at the Fairchild factory in Longueuil, caused a sensation. Its cargo hold was 21 feet in length; 5 feet, 3 inches wide; and 6 feet, 2.75 inches high. It had nine doors and hatches and, until the Second World War, registered as CF-ARM, it was the largest aircraft in Canada.

Richardson expected that it would substantially increase revenue by being able to fly more freight at lower prices and that all heavy hauling contracts would now come to Canadian Airways Ltd. But the BMW engine proved so unreliable that in its first year of service, the Junkers sat idle for 297 days. Engine problems continued over the years until, in 1937, it was re-engined with a Rolls Royce Buzzard engine. There were other problems not as easily corrected. The corrugated "Elektron" metal, which the pilots said gave the appearance of being in a Nissen hut on wings, cracked after a few hard landings. The double wing design (Junkers used full span ailerons and slotted flaps trailing behind and below the wing's leading edge) was an ice magnet. A few minutes flying collected ice in the gap between the wing and aileron and locked the controls. Like their Russian, Swedish, and German counterparts, the Canadian pilots learned to waggle the wings at intervals to prevent complete (and fatal) loss of control.

With its size and idiosyncrasies, memories of the Junkers lived on long after its use. Canadian Airway's engineers never forgot that the engine was cooled with ethylene glycol and that "it was hell of a job draining that thing, heating the glycol and putting it back." Others remembered CF-ARM as "an awkward old cow on the water because the pilot couldn't operate the throttle, watch where he was going, and operate the water rudders at the same time. She wasn't bad in the air, except she was slow. She had hinged flaps and floats as big as all outdoors! Standing beside those floats, the step was above you! Some of the places we went into didn't have docks that could handle her."

But CF-ARM survived to be sold to Canadian Pacific Airlines in 1941 and remained operational until 1943. With the Nazis in power, Richardson decided against buying any more Junkers, but those built in Spain flew there (and in the Spanish and Portuguese African colonies) long after the war had ended. Such was CF-ARM's importance to Canadian aviation history that the Western Canada Aviation Museum purchased a Spanish-built Ju 52/3m. It was flown over to Winnipeg and, with funding from the James Richardson Foundation, converted to single-engine status to resemble CF-ARM and put on display with the original Rolls Royce Buzzard engine.

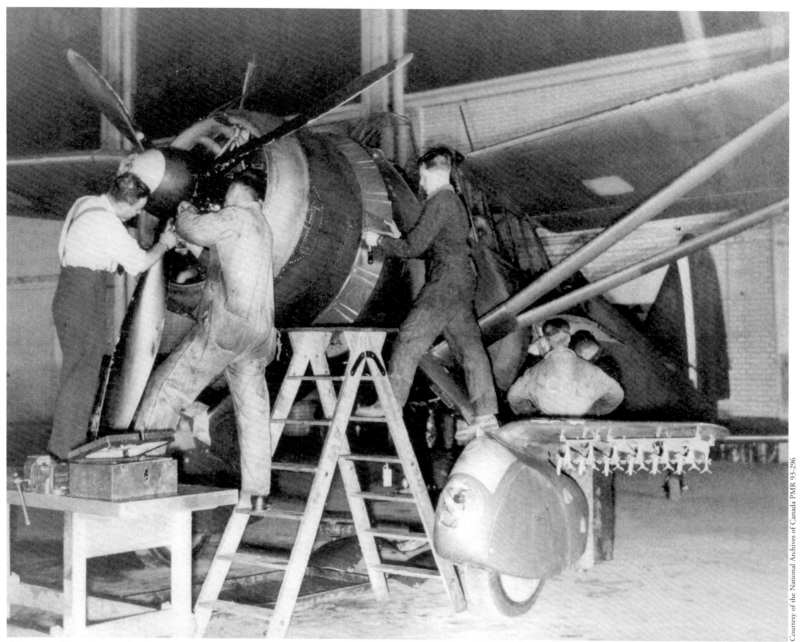

National Steel Car built 225 Lysanders at Malton. Note the bomb racks.

WESTLAND LYSANDER

The Lysander was a slow, gentle aircraft made for the British Army between the wars and named after the Spartan admiral who destroyed the Athenians in 405 B.C. (the army had a tradition of naming its reconnaissance aircraft after classical heroes). However, it was so slow and temperate that it more resembled Hermia's befuddled lover in Shakespeare's *A Midsummer Night's Dream.* To the troops in the field, the ungainly-looking aircraft was always the "Flying Carrot."

Even the Lysander's origins have a misty, sentimental appeal. Petter was a family-run company that built agricultural and dairy engines in Yeovil, Somerset during the Industrial Revolution. During the Great War, they asked the government how they could help (or cash in on) the war effort. With typical military logic, the Admiralty ordered Petter to build aircraft, and, renamed Westland Aircraft, it found itself in the aviation industry. In 1935, the Air Ministry wanted a two seat army cooperation aircraft to replace the aging Hawker Hector, and Westland's chief designer Arthur Davenport did a survey of the army squadrons to find out what the officers in the field wanted in an aircraft. The conclusion was that what was needed was a rugged aircraft that had STOL (short takeoff and landing) capabilities on any battlefield. It had to be able to fly very slowly and give the pilot excellent visibility from the cockpit.

Davenport took care of the wings by putting them through the top of the cockpit canopy and bracing them with struts. While this was revolutionary, his genius lay in the wing design. For the low-speed landing, the wing was thickest in the middle and thinnest at the root; the inboard leading edges he designed were the most unusual wings yet. It was also the first British aircraft to have trailing edge flaps and leading edge slats, which made the slow landing

Courtesy of the National Archives of Canada PA 167611

Westland Lysander in flight showing unusual wing design.

speed possible. The aircraft was first designed with retractable landing gear, but this was changed to fixed gear with streamlined fairings that would have landing lights and, if needed, a machine gun. The Air Ministry ordered 169 Lysanders. In 1939, Westland gave National Steel Car (NSC) of Hamilton, Ontario — the Canadian railway rolling stock manufacturers — the order to build them at Malton Airport in Toronto. NSC turned out 225 Lysanders before starting on Lancasters.

Seven Lysander squadrons accompanied the British Expeditionary Army to France in 1939, where, because of their slow speed and lack of armament, they fell easy prey to the Luftwaffe. When they were withdrawn to Britain, the Lysanders were assigned target tug and air sea rescue roles. Their real value was discovered in 1941, when the Special Operations Executive used Lysanders to drop agents and ammunition to the French Resistance, adding a ladder to the rear cockpit for the agents to climb in or out — and this is the role for which they would be immortalized. In Canada, the locally built Lysanders would be used for target towing and communications until sold for surplus after the war.

GRUMMAN GOBLIN

The American Grumman Goblin, like the Hawker Hind in Britain, could either been seen as the ultimate in biplane design or as an underbred monoplane. Both aircraft were the last biplanes built by their companies as they moved into the age of the monoplane fighter.

Leroy Randle Grumman was one of the earliest American naval pilots, training at Pensacola, Florida in 1918 before joining the tiny seaplane firm of Loening Aeronautical Engineering in New York. When that company was sold in 1929, Grumman mortgaged his house to begin his own in a Long Island garage. The onset of the Depression was hardly propitious for anyone in the aircraft industry, and Grumman might have disappeared from history had he not heard that the navy was looking for someone to design strong retractable landing wheels for its amphibious aircraft. Moving his tiny company to an empty building in Bethpage, Long Island, formerly owned by the Fairchild Aircraft Corporation, Grumman not only developed the landing gear but also built a biplane, XFF-1, to demonstrate it with. It had the traditional Grumman qualities (barrel shaped, stubby, powerful, and strong) and the navy liked it enough to place an order for many more. From these, Grumman developed his first naval shipboard fighter, the F2F, in 1932. It was a double or single seat biplane with a difference — while the landing gear did not quite retract into the fuselage, it did fold up into a boxy appendage beneath the fuselage. The F2F's 770 hp Wright Cyclone engine allowed it a creditable 220 miles per hour and range of 500 miles, and the United States Navy bought the aircraft to equip its land-based squadrons at Pensacola and Anacostia. But both Grumman and the navy realized that the day of the biplane was over, and no matter how nimble the F2F was, it was still a biplane. Besides, the navy was more

Grumman Goblin.

Courtesy of the Department of National Defence

interested in a monoplane called the Buffalo that was being offered by the Brewster Aeronautical Corporation. In 1936, Canadian Car & Foundry of Fort William, Ontario looked for an aircraft to build that would be within its capabilities and have some export potential. It chose the Grumman F2F, named it the Goblin, and constructed fifty-seven examples before realizing that this was a mistake. The RCAF were hardly enthused by another "hand-me-down" biplane and had no need of a naval fighter, but the company managed to export one Goblin each to Nicaragua and Japan. In what was thought of as an incredible marketing feat, forty Goblin fighters were sold to a Turkish company. When this was revealed as a front for the Spanish Republican air force, the Canadian government embargoed the sales, but by then it was too late. A few Goblins had already been shipped to the Spanish Civil War, where they ended up in the air forces of both combatants. But the experience of building the biplanes encouraged Canadian Car & Foundry to design and build Canada's first indigenous fighter aircraft, the FDB-1.

Early in the Second World War, the RCAF found itself in the position of beggars unable to be choosers, and the remaining Canadian Car Goblins, nicknamed "Pregnant Frogs" by the media, were taken on strength by 118 Squadron at Rockliffe, Ontario. With invasion scares on the east coast, the Goblins were moved to Dartmouth, Nova Scotia, becoming in those critical days Canada's only aerial defence in on its Atlantic coast. Fortunately for the Canadian east coast, the enemy never came by air, and in 1941, the Goblins were replaced by Kittyhawks. They were passed down to 123 Army Cooperation squadron, and disposed of the following year.

As for Leroy Grumman, the F2F (Goblin) enabled his company to develop other, more successful, fighters — like the Wildcat, the Avenger, and the Hellcat — in time to defeat the Japanese. Grumman died in 1982 as his company diversified away from naval aircraft to executive jets. Eight years after his death, it was bought by Northrop.

.

Courtesy of the National Archives of Canada PL 2081

Fairey Battle.

FAIREY BATTLE

The only thing bellicose about the Battle was its name. The Fairey Aviation Company had a reputation in the 1930s of building aircraft that were obsolete very soon after the first flight. One of the reasons it was still in business was that the Fleet Air Arm was its prime contractor, and it was not thought likely to encounter modern enemy fighter aircraft far out at sea. The Battle was an exception to this on two counts: it was not for the navy, and, when commissioned in 1933, it really was a state-of-the-art light bomber. The contract to mass produce it was given out to the "shadow factories," especially Austin Motors, which continued to build them long after Battles were known as "flying coffins."

Launched on March 10, 1936, at a time when the overall design of fighter and bomber aircraft had improved little since the Great War, the Fairey bomber had a closed cockpit, retractable landing gear, and the latest Merlin engine. In a day of biplanes, with their jungles of wires, the Battle was a svelte, low-wing monoplane that completely enclosed its crew of three (pilot, navigator/wireless operator, and rear gunner) and its bomb load as well. The Merlin 1 was enough to get a speed of 250 miles per hour, which made it faster than the RAF's Hawker Fury, and it had a range of over 1,000 miles. The only armament was a single Vickers drum fed machine gun in the rear and one .303-inch forward — hardly protection against attacking fighters, but then, none were expected. The glass house canopy was comfortable for the pilot and gunner, while the navigator/bombardier was buried inside the fuselage. Strangely, while the Merlin was in its streamlined cowling, the propeller didn't have a spinner, and the variable pitch gear was exposed. Fairey had also pioneered the retractable undercarriage that was indicative of the Battle, and pilots were so unused to it that the aircraft came with a klaxon to warn them to use it. Crews thought it easy to fly, if a little sluggish to the controls, but as long as one remembered to put the landing gear down — and a klaxon warned you to — it was easy to land. All in all, the Battle was the ideal light day bomber for 1933, a time when the RAF patrolled territories in the Middle East and a few bombs could discourage even the most warlike tribesmen. The philosophy of a light bomber that relied on speed rather than armament was copied by the new Luftwaffe, which equipped itself with the Heinkel H 70. Fairey built a thousand Battles, and, after the Munich Crisis, the contract was spread out to "shadow factories" for more; Austin Motors at Longbridge built 1,029 more.

Unfortunately, by the time the Battle equipped front line squadrons, fighter technology had taken a quantum leap. The Hurricane, Spitfire, and Me 109 all appeared by 1938, and although the later Battles were re-engined with the more powerful Merlin III, they were too slow and unprotected. They had neither the range for a really long-range bomber nor the speed for a fighter. Nevertheless, so short of monoplanes was the RAF that ten Battle squadrons were sent over to France with the British Expeditionary Force. Here they proved to be easy quarry for the Bf 109s and German anti-aircraft gunners, but somehow on September 20, 1939, when a Battle squadron (88) was sent to reconnoitre the Siegfried Line, one rear gunner managed to shoot down a Bf 109, the first "kill" of the war. However, throughout the "phoney war," Battles were destroyed by the Luftwaffe — usually from the rear, as their lone Vickers made them easy targets. What sealed the aircraft's fate was the German blitzkrieg of May 10, 1940. All RAF squadrons were ordered to take off and slow the advance, and the Battles were in the air constantly, flying low bombing runs from 250 feet. Their fuel tanks were not self-sealing, and without amour plate, they were vulnerable to even light machine gun fire. On that day alone, of the thirty-two sent up, thirteen were lost and the remainder were damaged. The next day, May 11, eight were sent to bomb the advancing German columns; one returned. With the panzers pouring over the canal bridges at Maastricht on the Belgian/Dutch border, the order was given to 12 Squadron to select volunteer crews for six Battles to bomb the bridges. All the pilots knew by now that they were going to certain death. In a replay of the charge of the Light Brigade, five aircraft took off, made their slow and low bombing runs into the crossfire of the waiting flak units, and none returned. The crew of one Battle was awarded the Victoria Cross, posthumously. Two days later, the remaining sixty-three available Battles were sent out, and when only twenty-eight made it back, the RAF withdrew the squadrons, and the survivors were sent home. After that, Battles were only used at night as target tugs, coastal patrol, and trainers. Battles were ideal for the Commonwealth Air Training Plan, and 739 were sent to Canada for that purpose. But so inadequate was Canada's home defence against prowling German submarines in the Gulf of St. Lawrence that Battles from No. 9 Bombing and Gunnery school were armed with depth charges to patrol over water — a mission that in an aircraft as poor as the Battle was suicidal.

FAIREY BATTLE

Today, only two Battle aircraft remain. In June 1940, convinced that the Germans would invade Iceland, the RAF re-equipped 98 Squadron, which had just limped home from France, with new Battles and sent it to Kaldadarnes, an airfield outside Reykjavik. One of those that crashed in the country's centre was salvaged in 1972 and rebuilt to be put on display at the RAF Museum in Hendon. The other is in the National Aviation Museum in Ottawa.

Beechcraft Expeditor. CT-33 and CT-128 Expeditor.

BEECHCRAFT EXPEDITOR

"Bug Smasher," "Wichita Wobbler," or "Shakey Jakes" (because of its Jacobs engines), the Expeditor had many well-earned nicknames. But it remained in production almost without modification from 1937 to 1969, an elderly workhorse that had gained the respect of civilian and military alike.

Walter Beech had built aircraft with Clyde Cessna and Lloyd Stearman before breaking away in April 1932 to form his own company in Wichita, Kansas. The height of the Depression was hardly a time to offer the public a personal plane. But Walter held that aircraft should not cost more than a family sedan, and he and his wife Olive Ann found the capital and personnel to turn out a five place staggerwing biplane that made the company's fortune. Until he died in 1950, Beech continued to build reliable, affordable light aircraft, one of which was the Model 18.

The twin-engined Beech 18 first flew in 1937 and was thrown into competition for the feeder market with the Lockheed 12 and Barkley Grow. What stimulated Beechcraft was the coming war, when the United States military chose the 18 to be a communications, navigational, and gunnery training aircraft, designating the military version the C-45. Under the Lend Lease policy, 236 Beech 18s were supplied to the RAF, the Royal Navy, and the RCAF, where they were known as Expeditors. On May 8, 1940, Canadian Airways acquired three Beechcraft 18s (CF-BQG, -BQH, and -BQQ) at thirty-three thousand dollars each for use on their Maritime route before selling them to Canadian Pacific Airlines in 1941.

The Royal Canadian Mounted Police bought the first of five Expeditors in 1946 for use in transport, search and rescue, anti-smuggling, and mercy flights, retiring the last in 1973. But the main users were the air forces around the world; after the war, the RCAF operated eighty Beech 18s. The maintenance contract for the RCAF fleet was given to MacDonald Brothers Aircraft of Winnipeg — Walter Beech and Grant MacDonald had been pre-war friends. In 1950, Beechcraft offered the RCAF the D18 version as a navigational trainer; three hundred were bought, designated the Expeditor 3N, all to be serviced at MacDonald Brothers. An offshoot of the Beech/MacDonald relationship was the arming of Beech 18s in 1952, when the Chilean air force asked to buy a number of D18s as ground attack fighters. Beechcraft, which initially did not wish to be involved, provided the Canadian company with salvaged fuselages, and MacDonald Brothers installed in each a pair of fifty-calibre machine guns in the nose, bomb racks, and provisions for firing rockets. So successful were these that Beechcraft then began marketing its own Mentors around South America as cheap ground attack aircraft. MacDonald Brothers continued to service the RCAF's Beech 18s until 1968. By then, the Canadian Armed Forces were using another Beechcraft, the Musketeer (CT-134), as primary trainers. The connection with Canada continues today; in 1980, Beechcraft became a wholly owned subsidiary of the Raytheon Company, which makes the Harvard II, the current primary trainer of the CAF.

Expeditor "CF-BQQ" in Canadian Airways service, allowing the crew to wear uniforms for the first time.

FAIREY ALBACORE

The Albacore was a good aircraft that had the misfortune to be overshadowed by its predecessor, the Fairey Swordfish. This was especially so after the Swordfish's successes in the raid at Taranto and the hunt for the *Bismarck*, after which the Albacore was relegated to more mundane duties. Nevertheless, in that capacity it ranged far across the world — from the Middle East to Malaya, from the Arctic to South Africa. Planned as an heir to the Fairey Swordfish, the prototype Albacore flew on December 12, 1938. As a biplane, at first glance it seemed to be hardly an improvement on the antiquated Swordfish; both had metal monocoque fuselages with fabric-covered wings. But unlike that of the Swordfish, the Albacore's fortunate crew of three worked in an enclosed, heated cabin. It also had almost double the range of the older aircraft at 932 miles, while carrying the same 1,610-pound torpedo or four 500-pound bombs. But its armament, like the Swordfish's, was pitiful: the starboard wing of the Albacore housed a .303-inch machine gun, and the rear cockpit held twin Vickers machine guns. The wise Albacore pilot put his faith instead in the aircraft's manoeuvrability. The Royal Navy received its first Albacore on March 15, 1940, and the aircraft joined Swordfish and Fulmar squadrons on the carriers HMS *Formidable*, *Illustrious*, and *Victorious*. But although it fought well in the Battle of Cape Matapan — the first sea battle in which aircraft played a decisive role — the Albacore was never destined to achieve its potential. Not only was the Swordfish indestructible, but the Fleet Air Arm was soon receiving modern American torpedo bombers like the Avenger.

But Fairey had the contract to build Albacores at its Hayes plant, and the production line could not be turned off so quickly, so eight hundred were made. Most were immediately relegated to rear guard squadrons, coastal defence, mine

Fairey Albacore.

Courtesy of the Department of National Defence

laying, and target marking. They flew patrols in East Africa, Malta, and the Middle East (the last one in 1946 at Aden). Others went to squadrons in Trinidad and Coimbatore, India. There were moments of battle, like January 26, 1942, when an Albacore from 36 Squadron was shot down attacking Japanese naval vessels off Endau, Malaya. Called the "Applecore" by its crews, the aircraft's last major victory took place on May 24, 1944, when, while on patrol over the Channel just before D-Day, one sank a German torpedo boat before it could disrupt the Allied preparations. The

Albacore landing on deck of aircraft carrier.

Royal Canadian Navy and the RCAF were saddled with Albacores in May 1943, the biplane serving in the RCAF until 1949. It has the distinction of being the last biplane used by the RCAF, and the only one used overseas.

Half a century later, the Albacore is still overshadowed by its predecessor. The single Albacore remaining today resides in the Fleet Air Arm Museum, Ilchester, Britain on static display. But one Swordfish is still kept airworthy by the Royal Navy at RNAS Yeovilton, and it will always be associated with the raid at Taranto.

Albacore wings folded.

Boeing 247.

THE BOEING 247

It was on April 2, 1933, that the United States aviation industry took the lead. Until then, the Europeans — Fokker, Junkers, and De Havilland — had dominated the design of commercial aircraft. The only American airliners of note, the Ford Trimotor and the Boeing 80, were clumsy hulks, adequate but hardly inspiring. The Europeans, with their state subsidies and imperial routes, had forged ahead, leading with long-range, four-engined airliners like the Armstrong Whitworth Atalanta and the Handley Page 42. The United States had been restrained in their aeronautical design until this Sunday morning, when the public had been invited to Seattle Airport to see the local manufacturer's latest product. It was a time of great expectations in the United States: the country had just been told by Franklin Delano Roosevelt, the new president, that they had nothing to fear but fear itself. Certainly the Boeing 247 aircraft on display outside the United Air Lines hangar was different from any other that anyone had ever seen.

Gone were the double wings, struts, and wires associated with biplanes, the triple engines (with one in the nose) necessary to power the aircraft over the Rockies, the corrugated sides and massive fixed undercarriage wheels of contemporary airliners. Instead, here was a smooth, monocoque shape with the internally braced wings cradling two Pratt & Whitney Wasp engines, themselves wrapped in streamlined cowlings. When the publicity handouts from Boeing/United Air Lines (it was the same company) claimed that the 247 could carry 10 passengers over 485 miles 50 percent faster than the Ford Trimotor, one could believe it. The day of the trimotor and the biplane was over. Rather than the fixed landing gear that all aircraft then sat on, its wheels retracted into the wings. For those allowed to peek into the aircraft, even greater delights awaited. What would they have noticed first — the lavatory, the indi-

Boeing 247 in flight.

vidual reading lights and air vents over each seat, the seats no longer made of wicker but upholstered as in a Pullman railcar? Rather than flying the aircraft while exposed to the elements, the pilot and co-pilot now enjoyed an enclosed, heated cockpit with the latest in two-way radio. In the coming months, those United Air Lines passengers who were fortunate enough to fly the 247 realized that the days of enduring engine noise, freezing temperatures, and airsickness due to seesaw rides were behind them. The modern era in air travel had arrived.

The Boeing 247 had no single parent, owing its birth to a whole family: Boeing Air Transport (later United Air Lines); Pratt & Whitney, for the Wasp engines; the metallurgists who developed Duralumin; Hamilton Standard, for the variable pitch propellers; and, most importantly, the farsighted United States Postmaster General Walter Folger Brown, who had forced the airlines to look beyond their airmail subsidies and carry passengers. For a brief moment, the 247 put United Air Lines at the forefront of air travel. Air travel wasn't only for businessmen — the ads proclaimed that even women could now fly. It allowed catering to passengers to become a whole industry. Stewardesses were invented by United Air Lines for the 247, as was airline food, carry-on baggage, airport limousines, and aviation advertising. As the stewardesses had to be trained nurses, for the very first time children, too, were encouraged to fly.

The 247 had its faults: the publicity did not mention the main spar that ran through the cabin, or the forward sloping windscreen that reflected landing lights and disoriented the pilots, or the two 247s that crashed because the rudder controls were too heavy. But what finally killed the forerunner of all modern airliners was its manufacturer's own shortsightedness. As Boeing made the plane exclusively for United Air Lines, Jack Frye, president of Trans World Airlines, went to Donald Douglas and asked him to build the DC-1. When that flew on July 1, 1933, it was the beginning of the end for the 247 — its day in the sun was over. In 1934, when a KLM DC-2 beat out a Boeing 247D in the MacRobertson Air Race between Britain and Australia, Boeing was already working on its Stratoliner and B-17 bomber.

Five Boeing 247Ds served with the RCAF from 1940 to 1942 in the transport and target-towing role, mainly with 121 Composite Squadron, Dartmouth, Nova Scotia. Then they were sold to Canadian Pacific Airlines for use on their northern and Quebec networks. The only Boeing 247D on display in Canada today is c/n 1699, which was in RCAF service as CF-BQS and then lent to Canadian Pacific Airlines as CF-BVX. After the war, it returned to the United States to be used as an executive transport and later as a crop duster. In 1956, a Calgary oil company needed an aircraft to service its northern drilling rigs and bought the 247D, which was re-registered as CF-JRQ. It was finally retired in 1967 and donated to the National Aviation Museum.

Avro Anson.

AVRO ANSON

"Faithful Annie" began as a requirement by Imperial Airways in 1933 for a fast, low-wing airliner. Whatever aircraft was chosen, it was to fly six passengers from Croydon to Brindisi, Italy and connect there with its seaplane service to the Middle East. Soon to make his mark with the Lancaster, Roy Chadwick designed the Avro model 652 for the purpose, and two were bought by Imperial Airways for the route. After that, the aircraft might have vanished into aeronautical history had not the Air Ministry, looking for a coastal reconnaissance landplane, placed an order for 174 copies of the 652. The little airliner's future was assured, and the Anson was born.

It is difficult to comprehend now, but when introduced in March 1935, into an air force used to biplanes with open cockpits and fixed landing gear, the Anson was so radical that a horn had to be installed to cut down on the number of belly landings. The flight crews of all the services — the Royal Air Force, the Royal Navy, and Coastal Command — loved Avro's modifications: the new "glasshouse" of huge rectangular windows for better visibility, the Armstrong Whitworth manually operated gun turret, the improved Cheetah engines, and the trimming gear and rudder control for cruising stability. The Anson was ideally suited for the coastal patrol task in 1939 — just as another former airliner, the Lockheed Hudson, appeared in military guise and rendered it obsolete. Until the American aircraft could be supplied in quantity, the Anson continued to hold the line — in that desperate summer of 1940, when Britain feared invasion, Ansons from 500 Squadron patrolled the Straits of Dover, fitted with downward-firing cannon.

The aircraft's greatest hour was to come, not in England, but in the far-off dominions. While Ansons had been used as pre-war bomber trainers in Britain, it was the signing of the Commonwealth Air Training Plan in Ottawa on December 18, 1939, that elevated them to the mass trainer role. They were shipped to flying training schools in Canada, Australia, New Zealand, South Africa, and Rhodesia. Canada received 1,528 Mk I Ansons; many were re-engined with the American 330 hp Jacobs (Mk IIIs) — distinguishable by their smooth cowlings — and later with Wright Whirlwinds and Pratt & Whitney Wasps. A new crown corporation, Federal Aircraft Ltd., was set up to supervise and coordinate Anson production across Canada. Among the companies involved were Boeing in Vancouver, which made the wing spars, and National Steel Car in Toronto and Macdonald Brothers in Winnipeg for the fuselages. To cope with the Canadian winter, the National Steel Car Company modified the British Ansons by removing the gun turrets, adding cockpit and carburetor heating, glazing the large windows, and panelling the interior. In total, the Canadians built 2,882 Ansons (Mk II, Mk V, and Mk VI), "Canadianizing" them with Jacobs engines. The Mark Vs were given the distinctive pert "Vidal" nose, and some were fitted with dual controls. To the relief of the poor flight engineers, who had to turn the crank 174 times to bring the wheels up, a hydraulic crank was also installed. In 1943, the United States Army Air Force was short of "bombardier" trainers and communications aircraft, so in a reverse "Lend-Lease," fifty RCAF Ansons, redesignated as AT-20s, were sent to the United States. In post-war Canada, the locally built Ansons, nicknamed "Bamboo Bombers," were the basis of many start-up airlines; the Department of Transport operated eight Mk Vs until 1956. One of the wartime Anson test pilots at Winnipeg was Roy Brown, who, with Milt Ashton, would begin Central Northern Airways (later Transair) using those aircraft. The Anson also proved to be "a good little earner" for Avro after the war, especially while that company wasted millions of pounds on its ungainly Tudor airliners and experimental Model 720 jet fighters. Production only stopped in 1952, when the company geared up for its successor, the 748, which turned out to be the last airliner built by Avro. The Royal Air Force flew its last Anson in 1968, more than three decades after receiving the first — "Faithful Annie" indeed.

LOCKHEED ELECTRA

In 1933, Robert Gross, chairman of the Lockheed Aircraft Corporation, realized that his company did not have the resources to match Boeing's 247 or Douglas and its DC-2. He decided instead to corner the regional aircraft market with a series of small airliners, the first of which was the Electra 10A. An outstanding aircraft in its own right, the 10A would bring modern commercial aviation to Canada.

Gross attracted to Lockheed talented aeronautical designers like Hall Hibberd, Richard Von Hake, Lloyd Stearman, and Clarence "Kelly" Johnson, and the Electra was their creation. The cantilever low-wing monoplane had two Pratt & Whitney Wasp Jr. 400 hp engines and a retractable undercarriage, and was equipped to carry a crew of two with ten passengers. Named Electra as part of the Lockheed "star" series, it was test-flown by Marshal Headle on February 23, 1934, only seven months after the first flight of the DC-1. Like the DC-1 and Boeing 247, the Electra was to have a single vertical fin, until Kelly Johnson determined that a rudder behind each engine increased control and stability, and the prototype Electra appeared in the twin fin configuration. The 148 Electras built were faster than the 247 and more economical than the DC-2 and sold well with airlines in the United States and elsewhere. Two would achieve fame: one British Airways 10A would take Prime Minister Neville Chamberlain to meet with Adolf Hitler; while another (the model 10E) equally ill-fated, would be used by Amelia Earhart on her doomed circumnavigation of the world.

Five Electras came to Canada, revolutionizing commercial aviation in a country where the biplane and the fabric-covered Fokker were still king. The first, CF-AZY, arrived at Vancouver Airport on August 4, 1936, having been

Lockheed Electra on the cross-Canada tour at Malton. Note DC-8 and Viscount in background.

bought by aviation entrepreneur James Richardson to influence the federal government into granting his Canadian Airways the trans-Canada air mail contract. On August 21, it was joined by the Electra CF-BAF, the second Canadian Airways purchase. When the federal government decided to start its own company, Trans Canada Airlines, both aircraft were sold to it — CF-BAF for pilot training, and CF-AZY for the Vancouver to Seattle run. So captivated was C.D. Howe, the minister responsible for TCA, with the Electra that in October 1937 he had three more purchased: CF-TCA, CF-TCB, and CF-TCC. They would go down in Canadian aviation history as the "Five Sisters" that established the early TCA network of routes. As the first all-metal twin-engined aircraft to land at many of the new airports along the Trans Canada Airway, they caused a sensation. The first generation of TCA pilots, some of whom would three decades later retire flying DC-8 jets, "cut their teeth" on them. The Electra would soon be replaced by Lockheed's Super Electras, and, in 1939, the RCAF would purchase the five from TCA for use as communications and training aircraft. While in RCAF service, CF-TCB would be destroyed by fire, but the other two (CF-TCC and CF-TCA) survived. In 1945, War Assets Disposal sold CF-TCA to the Thunder Bay Flying Club, which resold it to Wisconsin Central Airways. As NC 79237, it became an executive transport for the Florida company Bankers Life and Casualty Co., and would later be passed on to International Air Services, Lantona State Airlines, Florida, and Great Lakes Airmotive, Willow Run, Michigan. Eventually, the by now much abused Electra was crash-landed on her belly at Willow Run Airport and sold as scrap for five hundred dollars to engineer Lee Koepke. Koepke spent two years refitting her for a round-the-world flight with Ann Pellegreno in 1967 to commemorate Amelia Earhart's previous attempt. One of their crew was navigator William Pohemus, a former Air Canada navigating officer. On their return, Pohemus informed Air Canada that this Electra, now NC 79237, had once been CF-TCA. On March 11,

Ken Leigh Collection

1968, Air Canada bought and donated it to the National Aviation Museum at Rockliffe, Ontario. Repainted and re-registered as CF-TCA, she was flown to her final resting place by Air Canada Captain A.W. Ross and Vice President Herb Seagrim, who had first piloted her in 1937. As CF-TCA, the Electra took one last curtain call at the Air Canada pavilion at Expo 86. The other surviving 10A,

Pushing TCA Lockheed Electra out on its anniversary flight.

CF-TCC, was transferred to the DOT after the war and later purchased by Matane Air Services of Quebec to fly its Rimouski to Mont Joli to Sept Isles route.

In the summer of 1962, CF-TCC was borrowed by TCA staff to commemorate the twenty-fifth anniversary of the airline's first scheduled passenger flight. Gleefully, the old timers at all TCA stations volunteered to take part in the remembrance of the company's first flights. They decided to strip it of various layers of paint and return the Electra to the original metallic colour and then fly it across Canada, leaving Halifax on August 22 and arriving in Vancouver on August 29. It was to be flown by seven of TCA's first pilots: Captain Lindsay Rood and Captain Walt Fowler (Halifax-Moncton-Montreal); Senior Vice President Herb Seagrim (Montreal-Ottawa-Toronto); Captain Jock Barclay, flight simulator instructor (Toronto–North Bay–Sault Ste. Marie–Port Arthur/Fort William–Winnipeg); Captain George Lothian, superintendent of flying (Winnipeg-Seattle, with Captain René Giguere, flight operations manager Winnipeg as copilot Winnipeg-Regina-Saskatoon-Edmonton); Captain Art Rankin, flight operations manager Vancouver (Edmonton-Calgary-Lethbridge-Vancouver); and finally E.P "Billy" Wells, station operations manager (Vancouver-Seattle).

On August 14, 1962, two of the airline's icons, George Lothian and Al Hunt, ferried the Electra to the Air Canada Dorval base. As it was not possible to return it to its original metallic finish in the time available, CF-TCC was instead painted white and the names of the cities it was to stop at listed on one side. Everywhere the little aircraft landed, crowds came out to greet it. One of the passengers commented later that it was as if the Wright brothers had arrived.

On successful completion of the anniversary odyssey, the Electra was returned to Matane Air Services, and in 1965, it was sold in the United States. Ten years later, a former Air Canada pilot saw her in United States Army Air Corps colours performing for the Confederate Air Force in Harlingen, Texas. The old airliner was purchased by Air Canada in 1983 and refitted as CF-TCC for the airline's Golden Anniversary celebrations in 1986. She currently winters at the Western Canada Aviation Museum in Winnipeg while spending the summers on national goodwill tours for the airline.

Courtesy of the National Archives of Canada PA 124222

The first Lethbridge-Edmonton air mail flight. Posing with 10A CF-TCB are (l to r), W.R. "Wop" May, A.N. "Westy" Westerguard, C.M.C. "Con" Farrell, Sheldon Luck, Rudy Huess, Jock Currie, T.G.M. Stephens, airport manager Jimmy Bell, Jack Moar, and Harry Hayter.

DOUGLAS DAKOTA

The DST, DC-3, Dakota, C-47, C-49, Gooneybird, Li-2, Pionair, C-117, R4D-3, Skytrain, AC-47D — whatever the name, whatever the engines (Wright, Pratt & Whitney, Rolls Royce Dart, Conroy Turbo), civilization on this planet owes a heavy debt to this aircraft. From the first flight of the DC-1, on July 1, 1933, to the present day, it has flown millions of human beings honestly, scrupulously — and splendidly. It was not just the first aircraft to make its operators a profit, it was the aircraft that gave the world wings.

The DC-3's conception is better known than that of any other aircraft. Stymied by not being able to get any of the 247 airliners from Boeing, on August 2, 1932, TWA's chief of operations, Jack Frye, wrote the famous letter to Donald Douglas, Sr. asking him to build a rival. If the family of Douglas Commercial (DC) airliners has a single father, it is Arthur Raymond. Born in 1899, Raymond was the son of a California hotelier. As a teenager, Arthur went up in a dirigible with the famous balloonist Roy Knabenshue and never forgot the experience. When his father was bankrupted by the Depression, Raymond went to work for Donald Douglas and, besides his masterpiece, the DC-3, also designed the DC-4, DC-6, DC-7, and DC-8, ultimately playing a key role in the NASA Gemini and Apollo missions. Donald Douglas was also fortunate in that he could call on the talented Jack Northrop to design the DC-3's cantilever multicellular wings, and that the new high tensile aluminum alloy Alclad had just been developed.

From the single DC-1 made, Douglas progressed to the DC-2, which was bought by TWA and made by Fokker under licence in the Netherlands for KLM; the first one, "Univer," won second place in the Britain to Australia MacRobertson Air Race of 1934 — beating out a Boeing 247. That year, the president of American Airlines, C.R.

Douglas Dakota.

Courtesy of the National Archives of Canada PCN68-191

Smith, wanted to replace his Curtiss Condor biplanes and go one up on TWA by being able to fly fourteen passengers across the country with as few stops as possible. Smith's crucial sales pitch was to have his passengers arrive refreshed at the other end. But rather than losing time overnighting at refuelling stops, they would sleep longer on board the aircraft. At first Douglas was hesitant; the aviation industry was still in the Depression, and while the DC-2 had been a success, airlines could not possibly absorb another airliner. But Smith obtained the financing for Douglas to expand the DC-2's fuselage and wings and bought twenty of the Douglas Sleeper Transports (DST), called the "American Eagle," which came with windows in the roof for the sleeping passengers.

Douglas kept the DST aircraft in production as a day version, the extra room allowing it to now seat twenty-one passengers, enough to cover the cost of the flight, and when twenty-eight or thirty-two were squeezed in, profits could be made for the first time in aviation. The same basic semi-monocoque Alclad fuselage was used, except now it was rounded in the cross section to allow for more seats. Fuel tank capacity was also increased from 510 gallons to 822 gallons in the DC-3. So adaptable and robust was the DC-3 airframe that, as with the military C-47, its range of 2,125 miles could be increased even further with an extra nine 100-gallon tanks in the fuselage (with the C-47 floatplane, 300 gallons could be carried in each float). No one seems to have thought much of these innovations, and on December 17, 1935, when the first DC-3 took off from Clover Field, Santa Monica, there was no photographer to record the event.

United Air Lines quickly sold off its Boeing 247s and bought the "day plane," as it was first called, putting it in service on January 19, 1937. As it was the only medium-range transport available, the United States Army Air Corps ordered a military version, the C-41, the first of many variants (the C-47, which first flew in January 1942, was the most numerous). Aptly called the "Skytrain," it could carry an amazing seventy-four GIs! The Douglas plant in Santa Monica built 579 DC-3s and 382 C-47s, but the two factories set up during the war in Long Beach and Oklahoma City turned out 4,285 and 5,400 respectively. The Japanese first bought 20 DC-3s and then built 487 as L2Ds under licence. The Russians sent aircraft designer Boris Lisunov to Santa Monica for two years to study Douglas production, and when he returned home they built two thousand DC-3s at Tashkent as Li-2s. By 1946, Douglas had manufactured 10,147 DC-3s/C-47s, thus ensuring the model's immortality — the last off the line for SABENA.

By then the aircraft was already a legend. Photo footage of them dropping paratroops at D-Day Arnhem or supplying the British Army in Burma became synonymous with Allied victory. The stories of DC-3s surviving collisions, flak, and Japanese Zeros are legion. The most incredible one concerns a Chinese DC-3 that lost a wing during a Japanese air raid. Because there was no DC-3 wing available, one from a DC-2 was fitted, and the aircraft flew out without ill effects. During the war, the RCAF received 570 Dakotas to be used by 435 and 436 Squadrons in Burma and 437 in Europe. By 1979, there were still nine of those aircraft in service, older than the pilots who flew them. The final two were redesignated CT-129s and used as electronic warfare trainers.

As the turboprop and jet age dawned, manufacturers fitted the DC-3 with experimental turboprops, jets, and even rocket power. During the Vietnam conflict, DC-3s were redesignated as AC-47 Gunships (called "Puff the Magic Dragon" by the GIs) armed with Vulcan miniguns that fired through the windows on the port side for ground strafing. In the end, the future of the DC-3 lies not with its structure or the availability of spare parts, but with the price of aviation fuel. Arthur Raymond, the man responsible for the immortal aircraft, died on March 22, 1999, two months short of his one hundredth birthday. On that date, it was estimated there were still two thousand of his DC-3s in the air.

Courtesy of the Department of National Defence

Canadian Forces Dakota in D-Day colours.

HAWKER HURRICANE

More has been written about the Hurricane than any other aircraft of World War II. It is generally accepted that it rather than the Spitfire won the Battle of Britain, outnumbering as it did all other British interceptor aircraft combined. Its creator, Sidney Camm, could never have visualized it being used in the Battle of the Atlantic, flying off aircraft carriers as Sea Hurricanes or as Hurricats from CAM (Catapult Aircraft Merchantmen) freighters. Hurricanes were flown on every front, as "tank busters" in North Africa and as jungle fighters at Kohima in Burma. They fought with every British ally and many potential allies, with the Red Air Force and, in the hands of the Finns, against them. Its uses were legion — the eight Browning guns were superseded with four 20-millimetre Hispano and later two 40-millimetre cannon. But before the Hurricane accomplished all of that, it already had two historic pre-war firsts: it was the first British and Canadian monoplane fighter and the first to exceeded three hundred miles per hour.

While famously remembered as a stable gun platform, the Hurricane's greatest merit was, in contrast with the thoroughbred Spitfire, its unsophisticated structure. Hawker had gained much experience with its previous fighters, the Hart and the Fury, and the Hurricane set a new standard in fighter technology. That this versatile aircraft was built at all was due to the brilliance and tenacity of the chief designer at Hawker, Sidney Camm.

As early as 1925, figuring that biplane design had reached its culmination, Camm had designed a monoplane fighter, but the hidebound Air Ministry held that monoplanes were for racing aircraft only and lacked the manoeuvrability of biplanes. Camm continued with development of his "Interceptor Monoplane," codenamed K5083, as a private venture just as Hawker purchased the Gloster Aircraft company plant. Thus, when Britain began rearming and officialdom now looked

Hawker Hurricanes.

favourably upon the K5083, it was already in the prototype stage and flew on November 6, 1935. Camm had combined the latest Rolls Royce P.V. 12 engine (soon to be famous as the Merlin) with the traditional Hawker fabric-covered fuselage of cross-braced tubular metal. Innovative too was the enclosed cockpit, the retractable landing gear, and the cantilever twin spar wing, thick enough to house eight American Browning .303-inch guns. That July, now named Hurricane, the fighter was displayed to great acclaim at the Hendon Air Show, and an order unprecedented in British history was given for six hundred of them. RAF squadrons started to receive their first monoplane fighters in December 1937, and because Hawker now had the extra factory space from Gloster, it could produce Hurricanes at a brisk pace, even exporting them to Turkey, Belgium, Poland, and Yugoslavia (the Yugoslavs fitting theirs with German Daimler Benz engines) before the war began.

On the day that war was declared, there were 315 Hurricanes in operational squadrons, with 107 in reserve, and, thanks to Camm's foresight, the country had a credible air defence. During the Battle of Britain, outclassed by the German Bf 109s, the Hurricanes were allotted the Luftwaffe bombers. Sturdy gun platforms they were, the tubular construction allowing them to take whatever punishment the Luftwaffe gunners sent them. Hurricanes could be easily repaired at forward airfields, and their strong, wide track undercarriage coped with rough landing areas; their numerical availability eventually won the battle.

The first Hurricanes to arrive in Canada were Mk Is, which were delivered in 1938 and assembled in Vancouver. Although one had the misfortune to crash into Grant McConachie's old Ford Trimotor at Sea Island Airport, enough were ready in June 1939 for No. 1 Squadron to become operational and provide an air escort for the Royal Tour of 1939. The Canadian Car & Foundry company had been allowed by the Air Ministry, on January 4, 1939, to build Hurricanes, but it would be a full year before the first was flown. On June 11, 1940, No. 1 Squadron was shipped overseas with its Hurricanes in time to take part in the Battle of Britain. The later Canadian-built Hurricanes were powered by Packard-built Merlins and designated Mk Xs (with the XII built in greatest numbers). While some were kept in Canada for defence and as trainers, in 1942, others were delivered by convoy to the Soviet Union and further afield to the Indian Air Force. Canadian-built Hurricanes were easily distinguished from their British brothers because their large American propellers could not accommodate the British spinners. In all, a total of 1,451 Hurricanes (Mk X, XI, and XII) were built at Canadian Car & Foundry, a small number of them twelve-gun Sea Hurricanes for use on the convoy escort carriers that moved between Newfoundland and Iceland. Interestingly, these took part in "Operation Torch," the Allied landings in Algeria in November 1942. Because the American gunners were unused to Hurricanes, the aircraft flying off the carriers HMS *Biter* and *Dasher* were painted in American markings. When Vichy French Dewoitines shot down an entire squadron of RN Fairey Albacores, the Sea Hurricanes went into action and shot down five of the French fighters.

When the last Hurricane flew on July 1944, a total of 14,333 had been built. Camm was knighted in 1955 and died on March 12, 1966. But by then he had developed a long line of fighters, culminating in the Hawker Hunter, another versatile fighter and heir to the Hurricane.

North American Harvard.

NORTH AMERICAN HARVARD

In its present incarnation, the Harvard trainer is a sleek turboprop built by Raytheon, one that could never be confused with the original aircraft of that name. That Harvard was synonymous with two things: the "Yellow Perils" of the Commonwealth Air Training Plan and the racket its engine made. An anachronism even when it was manufactured in the 1950s, for many who trained on it, the Harvard was the last "real" plane that the RCAF flew.

In December 1928, a number of companies connected with aircraft manufacturing merged as North American Aviation Inc. When the Roosevelt administration forced such holding companies to break up, part of North American relocated to California. The president, James Howard "Dutch" Kindelberger, realized that, without much experience in the industry, the company ought to concentrate on building a simple training aircraft. The Army Air Corps needed one, as did the British and French. In 1934, North American developed the NA-16, a two-seat, low-wing monoplane with fixed landing gear, at its Inglewood plant. When it won the Corps contract, this was designated as the BT-9 (for basic trainer). With the next model, the NA-26, North American wanted to cover all bases befitting a training aircraft, and a strong retractable undercarriage was added. For the export market, where it was sold as a light attack bomber, provision was made for two cowl and two wing-mounted guns and a four-hundred-pound bomb load. The aircraft was offered with either the Pratt & Whitney R-1340 Wasp or the heavier, more powerful Wright Cyclone R-1820 radial engine. The 1,455 fixed-gear NA-26s built in the United States, Japan, and Sweden were outnumbered by the 20,110 retractable-gear models built altogether. A good basic design, it was called Harvard in the United States, Canada, and Britain and Wirraway in Australia. It had cousins called Texans (built in Dallas of course) and Yales. Some

Texans were ferried through Canada to the Soviet Union, while the French Air Force bought the Yale in great numbers, calling it the "North" — 111 of which were in France when the Germans took over. The Luftwaffe recognized a good basic trainer, and the Yales flew in German markings until the very end of the war. The undelivered Norths were sent, still in their packing cases, to National Steel Car at Malton for use as intermediate trainers.

These were gratefully accepted in Canada, as even before the war had broken out, the absence of advanced air trainers for RCAF pilots was marked. The British Purchasing Commission ordered four hundred Harvard Mk Is for itself in 1938, and, while some did make it to Britain, for the RCAF this was a hat trick: the aircraft were in production, were paid for by the British, and could be supplied from the factory directly to an RCAF base. By July 1939, the first Harvards in the RAF training colour of basic yellow (with a black antiglare panel in front of the cockpit) were being ferried from Inglewood to Vancouver and then to Canadian training stations like Borden and Trenton. With the signing of the Commonwealth Air Training Plan (CATP) on December 17, 1939, the need for trainers went into high gear, and in the following month, January 1940, the Montreal bushplane manufacturer Noorduyen was awarded the contract to build Harvards at its Cartierville plant. Exactly one year later, the first Canadian-built Harvard was delivered with some ceremony to Rockliffe, Ontario on January 31, 1941. With its Cartierville neighbour Canadian Car & Foundry (CCF), Noorduyen became the main facility for the repair of Harvards, the winterization of the aircraft, and the experimentation with wooden rear fuselages.

The story of the Harvard in the CATP is well known, but its post-war service is less so. In 1946, CCF took over

Noorduyen's production and produced the Harvard 4. The Cold War was beginning, and manned interceptors like the Avro Arrow would require thousands of pilots. In 1946, home defence in the form of "weekend warrior" auxiliary squadrons became the trend, and when some were equipped with P-51D Mustangs, the well-matched Harvard came into its own once more. The exuberance

Two Harvards and a Mustang.

of the Harvard made it a natural for air show demonstration, and there are and have been several teams, starting with the "Easy Aces," who performed clover leafs and Cuban Eights at the Canadian National Exhibition in Toronto from 1952 to today. The best known are undoubtedly the Harvard 4s of the "Goldilocks," a tongue-in-cheek tribute to the RCAF's "Golden Hawks" Sabre team. When, in 1950, Canada became the home to another air training plan, in which the cadets were not restricted to the Commonwealth but came from all of the air forces of all the NATO nations, including the German Luftwaffe, Harvards were reused as flying classrooms.

But even such a beloved aircraft was replaceable, and when the RCAF received its first Canadair Tutor on October 29, 1963, it was the beginning of the end. The final RCAF Harvard sortie took place on May 21, 1965, and the trainer was sent to Crown Assets Disposal or to museums. But for a generation of air enthusiasts who never saw them in air force colours, the Harvard lives on ... in Hollywood movies and on television. Ironically, Harvards routinely appear disguised on the screen as Japanese Zeros and German Focke Wulf FW 190s. But then, as North American Aviation knew in 1934, this aircraft could do anything.

Lockheed Hudson.

Courtesy of the Department of National Defence

LOCKHEED HUDSON

In the early summer of 1938, the British Purchasing Commission was touring the United States, looking to buy a reconnaissance plane/light bomber. The shortcomings of the Avro Anson for coastal maritime patrol duties were increasingly obvious, and the Air Ministry saw the Lockheed Super Electras used by British Airways as a possible replacement.

The Super Electra, designated the Model 14-H2 by Lockheed, had been designed by Hall Hibbard and Clarence "Kelly" Johnson in late 1936. A twin-engined, twin-tailed monoplane with a retractable undercarriage and Fowler flaps, it could be ordered with a variety of Pratt & Whitney engines. The Super Electra first flew at Burbank on July 29, 1937, and received its Approved Type Certificate in November, going immediately to the "launch customer," Northwest Airlines. Unlike its predecessors, the Lockheed Model 10 and 12, the Super from its inception was designed to be more than a passenger plane. On July 10, 1938, the millionaire aviator Howard Hughes flew a Model 14 (N) powered by Wright Cyclone engines around the world in ninety-one hours and fourteen minutes. His initial choice had been a DC-3, but Hughes had changed his mind at the last minute and bought the Lockheed. It was considered a fast, record-breaking aircraft, and this was fortunate, as its seat-mile economics compared unfavourably with the Douglas DC-2.

Lockheed built 112 Model 14s but suffered when Northwest Airlines, after three crashes, lost confidence in the aircraft. While only the first crash was related to a design flaw in the 14's tail, Northwest quickly dumped its Lockheeds in 1939 for DC-3s. But the Super Electra sold well overseas. In Canada, on the recommendation of Minister of Transport C.D. Howe, Trans Canada Airlines (TCA) bought sixteen 14-Hs, to be delivered from June 1938 to August

1939, and became the manufacturer's best customer. Later deliveries had the more powerful Pratt & Whitney Twin Wasp engines, were "stretched" by five and a half feet to accommodate four more seats, and were known as the Lodestar (18-08). Even before the Second World War began, Lockheed was actively courting the British Purchasing Commission (soon to be followed by the French and Dutch), all of whom were looking to purchase a light bomber. Keeping the wing, tail, and engines of the 14, the design team hurriedly adapted its Super Electra to seat a crew of four, added nose and dorsal gun turrets and, where the cargo hold had been, a bomb bay. The British bought the new version, renamed the B14, but asked that the nose gun turret be eliminated, the nose made transparent, and the navigator shifted into it to double as a bombardier. Lockheed agreed to these modifications, and on June 23, 1938, the British ordered 250 B14s — christened the Hudson Mk I by the RAF — all to be delivered before December 1939. It was the first American-made aircraft to be used operationally by the RAF, and its purchase caused a controversy in Britain. In fact, the Super Electra, or Hudson, would find itself flying for both sides during the war. After buying 30

Courtesy of Ken Leigh collection

Lockheed Lodestar in Trans Canada Air Lines service.

of them in 1938, the Japanese aeronautical company Kawasaki Kokuki K.K. acquired the rights to build them, and 121 were manufactured in Japan. They were designated military transports and code-named "Thelma" by the Allies.

The first batch of RAF Hudsons arrived in Britain by sea, either shipped from Long Beach, California or flown to Floyd Bennett airfield in New York, where they were taken apart and put on board ship for Liverpool, to be reassembled at Speke Airport near the docks. The scheme worked well enough until Congress passed the Neutrality Act, which forbade the delivery of aircraft directly to the combatants. Concurrently, German submarines kept sinking the Hudson-carrying ships at an alarming rate, making the sea route too dangerous. Lockheed then had its factory crews fly the B14s as close to "belligerent" soil as possible, either at Pembina, North Dakota, where they were towed across the United States–Canada border, or directly to St. Hubert, Montreal. Of those 250 Hudsons, 28 were kept in Canada and were assigned to the RCAF. One of them from 31 OTU sank a German submarine off Nova Scotia on July 4, 1943.

There would be several later models of the Hudson, from Mk II to VI, and all had strengthened airframes. When President Roosevelt signed the Lend Lease Bill on March 11, 1941, the United States Army Air Corps could take charge of the Allied purchases, and the Model 14s ordered were designated A-28 and A-29 (depending on the engine and mission of the aircraft). Early in the war, the A-28 was used at naval air stations for harbour and convoy patrol, but the former airliner was woefully inadequate in Europe for the anti-shipping role that the RAF put it to. Its two fixed .303-inch machine guns firing forward and two in the turret were no match for marauding Luftwaffe Ju 88s, and the RAF and RCAF squadrons that endured the Hudson were pleased to receive Bristol Beaufighters in their place. After the war, Canadian Pacific Airlines used four Hudsons (CF-CRJ to CF-CRO) from 1946 to 1949, selling them when DC-3s became more plentiful.

Curtiss P-40 Kittyhawk.

THE CURTISS P-40 KITTYHAWK

Forever enshrined in popular memory as the shark-mouthed fighter aircraft used by General Claire Chennault's "Flying Tigers" in China, the P-40 Warhawk was destined to be the perennial bridesmaid in aerial warfare — flying in rear areas or less than glamorous campaigns. Known as the Tomahawk to the British, the fighter served valiantly in the early years of the war, holding the line until the Allies could field the P-51 Mustang and the P-38 Lightning.

The P-40 had been designed by Curtiss Wright's Dr. Donovan R. Berlin to improve on his earlier P-36 radial-engined air-cooled fighter. The little P-36 had no competition when it flew in 1935, and the army air corps bought it in large numbers. But by 1938, with the world re arming, American aircraft designs were rapidly being overtaken by European designs, especially in speed and manoeuvrability at high altitude. Pratt & Whitney and Wright had been producing radial air-cooled engines for the commercial market, and these were sufficient for carrier-borne aircraft that would never have to face German fighters. Thus, when Allison brought out its series of V-1710 inline liquid-cooled engines, American aircraft companies seized on them to power their latest designs — Bell put one in its P-39 Airacobra (behind the pilot), and Curtiss's Berlin modified a P-36 to use another V-1710. The first Curtiss XP-40 flew on October 14, 1938, but by then British and German aircraft design had raced so far ahead that it was already obsolete. However, the army air corps were so short of fighters that they ordered 540 P-40s from Curtiss for its front line squadrons, some of which were to meet the Japanese at Pearl Harbor. The French ordered 140 P-40s, which the British took over when Paris fell. The British then ordered their own P-40s, calling them Tomahawks. Part of the British order of P-40s was sent to China to equip the American Volunteer Group; there, they were known as Kittyhawks. If the RAF found them inferior to the Luftwaffe in speed and at high altitude in Europe, their ruggedness and versatility made them ideal as close air support and air defence for parts of the Empire. It was in North Africa that an Australian pilot from the RAF's 112 Squadron first painted fangs on the prominent air intakes of his Tomahawk.

Curtiss built 13,737 P-40s from 1939 to 1944, modifying the armament, changing the cockpit for better visibility, and adding armour plate and more powerful engines. In time, they served with the Russians, the Australians, the New Zealanders, the South Africans, the Turks, and the Canadians. The Japanese attack on Pearl Harbor on December 7, 1941, galvanized the RCAF into providing for some defence of the west coast, which until then had been guarded by obsolete Blackburn Sharks. On December 8, Kittyhawks from 111 Squadron based at Ottawa were rushed to Sea Island Airport, Vancouver and then later moved to Patricia Bay when more P-40s arrived. These fighters became the main defence of the Canadian west coast until 111 Squadron posted to Anchorage, Alaska, under Wing Commander Gordon McGregor of future TCA fame. In this theatre, Squadron Leader K.A. Boomer, flying a P-40, shot down a Japanese "Rufe" seaplane over Kiska on September 25, 1942 — the only air victory by the RCAF defence squadron.

Kittyhawks also played their part in the defence of the Canadian east coast during the anti-submarine campaign in the mouth of the St. Lawrence. When German submarines sank merchant ships off the Gaspé coast in July 1942, 130 Squadron's Kittyhawks, based at Mont Joli, provided air patrols until more suitable aircraft were available.

SUPERMARINE SPITFIRE

The air aces who flew her (Douglas Bader, Johnnie Johnson, Al Deer, Robert "Buck" McNair, and Gordon McGregor) all bring the magic of the aircraft's name to life. The Spitfire was the most famous, most beloved aircraft of World War II. It was never just an aircraft — it was as much a symbol of Britain as the Union Jack, Sir Winston Churchill, and the white cliffs of Dover. There might have been better performing fighters (the Focke Wulf 190 comes to mind). There might have been more aesthetically beautiful fighters, like the P-38 Lightning. And even in the Spitfire's finest hour, the Battle of Britain, the humble Hurricane was more numerous and shot down more of the enemy. Yet the Spitfire holds supreme in our psyche, with many still believing that it was this aircraft alone that saved Britain. Its image is one of a knight sallying forth to slay the dragon. It is the immortal, ultimate symbol of defiance. Even its designer, Reginald Mitchell, was glamourized when actor Leslie Howard played him in the movies. As the chief designer at the Supermarine Company, Mitchell realized that his 1925 F.7/30 racing design was a perfect vehicle for a future fighter. There is no doubt that when it first flew on March 5, 1936, the Spitfire's alloy-covered monocoque fuselage, the elliptical single spar wing, and the Rolls Royce Merlin II engine were technically innovative, and the experimental aircraft would have been a fine aerobatic plane credit. What gave the Spitfire its timeless appeal were the clean lines of its nose cowling. In an age of radial engine biplanes, Mitchell had adapted the asymmetry of a racing machine to a production aircraft. The radiator, the oil cooler, and the carburetor intake were all hidden away beneath the wings and under the center. Although he could not have known about the effects of Mach speed on wings, Mitchell had made the Spitfire's wing as thin as possible. This would allow for its agility and adaptability in future marks.

Supermarine Spitfire.

Mitchell was very ill, but when he died in June 1937 at the age of forty-two, it was with the satisfaction of having the Royal Air Force order three hundred of his aircraft. His successor at Supermarine, Joseph Smith, realized Mitchell's genius and exploited his design to the maximum. As a result, the aircraft remained in production for twelve years — from the day of the biplane to the advent of the early jets. Its versatility was unique, and it was used in low level "Rhubarbs" over France, in Soviet skies, and in the desert against Rommel. In an effort to improve its performance as a low-level fighter, its wings were clipped by about four feet. In the Seafire version, they were folded to fit beneath the deck of aircraft carriers. While not robust enough for repeated deck landings, and a lethal "ditcher" at sea, the Seafire did give the Fleet Air Arm a high-powered fighter aircraft.

The original wooden fixed-pitch two-blade propeller received more blades until there were five. In 1947, the Seafire was fitted with contra rotating propellers. The armament was increased from eight machine guns in 1937 to four 20-millimetre cannon in 1946. The aircraft took on different shapes. Tropical air filters were added to operate out over the Bay of Bengal and Australia, and the old flat canopy became a bubble shape (introduced, it was said, because the pilots from Canada and Australia were taller). It was painted pink in the high altitude version. It was fitted with floats. It dropped canisters of beer to thirsty troops off the D-Day beaches and dive-bombed V-2 sites around The Hague. The Spitfire XII adapted to the Griffon engine, and it had its cockpit pressurized and its wings stretched for the high altitude VI version. The long-range photo reconnaissance version, the X and XI, had a range of two thousand miles. Spitfires flew off the aircraft carrier HMS *Eagle* in March 1942 to save Malta. Griffon-powered Spitfires destroyed more than three hundred V-1 flying bombs. It shot down the first enemy aircraft of the war and, by the end, the first German jet, the Me 262. The Spitfire Mk XIX had a ceiling of 43,000 feet and speed of 460 miles per hour.

Courtesy of Public Archives PL 120-732

Not bad for an aircraft begun in 1937 with 34,000 feet and 355 miles per hour. Supermarine kept adapting and improving their aircraft well into 1946, with the Seafang and the Spiteful, both getting further away from the original beauty of design that Mitchell had planned. But by then the company was already working on its first jet, the Swift.

The immortal Spitfire.

What British ally didn't operate the Spitfire? It flew with thirty air forces on six continents. The Americans flew them in the Eagle Squadron before their country entered the war, and one thousand were delivered to the 8th Air Force when it did. After the war, the French took theirs to Indochina, the Dutch to the East Indies, and in the Middle East, Egyptian Spitfires fought Israeli.

Mitchell had wanted to call it the Shrew, preferring it to the name Spitfire. While the aircraft would have been as potent and aesthetically pleasing whatever the name, it will always be the magnificent Spitfire in history. It was the finest fighter aircraft that the British ever made — perhaps the finest of all time.

AVRO LANCASTER

When the Avro Manchester III first flew on January 9, 1941, the "Clearance to Fly" form stated that its name was Lancaster. Historians mark this as the very first time the name of what was to become the best bomber of the Second World War was ever used. The union of four Rolls Royce Merlin engines with the Manchester bomber airframe, the Lancaster owed its existence to Avro's chief designer, Roy Chadwick. Not only had he been primarily responsible for the Manchester airframe, but he had also worked his "old boys" connection with friends at Rolls Royce to hijack the four Merlin engines. There was still some official reproof about not proceeding through the proper channels for the Merlins, but in June 1941, the Air Ministry placed an order for 450 Lancaster bombers, and Avro knew it had a winner.

The National Steel Car Co. of Hamilton, Ontario had only made railway rolling stock when it was awarded a contract in 1938 to manufacture Westland Lysanders at the newly opened Malton Airport in Toronto. They had progressed to the twin-engined Avro Anson when, in December 1941, C.D. Howe nationalized the Malton plant for a program of the highest wartime priority. Now named Victory Aircraft Ltd., it was to build Lancaster bombers. Avro sent over a Lancaster R-5727 to be used as a pattern aircraft, and in an incredibly short time, on August 1, 1943, the first Canadian-built Lancaster KB700 was rolled out. Four hundred twenty-nine such bombers would follow, all except eight destined for the RCAF.

These eight became Canada's first civil transoceanic airliners, the idea originating out of Howe's despair at finding a suitable aircraft with which to transport the mail to Canadian troops in Europe. In March 1943, when the pattern

"Lanky the Lancaster" at Dorval, with a Canadian Colonial DC-3 in the background.

Lancaster was declared surplus, the Department of Transport used it to fly heavy material from Moncton to Goose Bay, Labrador for the construction of an airport. Then, with Avro's consent, Howe commandeered R-5727 for his own purpose. Typically, as with anything connected with the minister, things happened quickly after that. On April 2, 1943, Trans Canada Airlines was given the right operate a non-commercial air service between Montreal and Prestwick, Scotland, which for legal and diplomatic reasons was to be under a separate company, the Canadian Government Trans Atlantic Air Service (CGTAS). The pattern Lancaster was registered as CF-CMS and loaded with twenty-six hundred pounds of mail and three official passengers. On July 22, 1943, the first CGTAS flight took place. The crew (Captain R.F. George, First Officer A. Rankin, and Radio Officer G. Nettleton) were all TCA staff, except for Navigator J. Gilmore, who was borrowed from the RCAF. Twelve hours and twenty-six minutes after takeoff, CF-CMS landed at Prestwick — the first Canadian commercial aircraft to cross the ocean. The pattern Lancaster was joined by four Victory Aircraft–built Lancasters in 1943, and by another four in 1945 — all registered CF-CMS to CF-CMZ and then CF-CNA.

An ungainly looking, gangly aircraft on the ground, the CGTAS Lancaster could never be mistaken for the svelte United Air Force DC-4s or the purposeful BOAC Liberators with which it competed on the Atlantic run. The converted bomber was a thin, rawboned, "tail-dragger," rearing up on giant wheels and lacking all passenger amenities such as heating and insulation. It was difficult to load mailbags into and didn't fit into the hangars for servicing. Yet TCA personnel were intensely proud of Canada's first transoceanic passenger transport and referred to it affectionately (if unofficially) as "Lanky the Lancaster." The TCA crews knew that they were flying into a war zone and prided themselves on CGTAS being "an all Canadian show...and that all the planes but the first one were made in Toronto."

There were casualties: CF-CMU vanished over the Atlantic on December 28, 1944, and the original aircraft CF-CMS caught fire while on a conversion flight on June 1, 1945, crashing into a farm just outside Dorval Airport. When the war ended, CGTAS took on a commercial form, selling tickets and advertising for passengers. Because it could fit only ten passengers into the hold, and the International Air Transport Association had fixed the per person fare for crossing the Atlantic at $375 each way, it would be impossible for TCA to make any prof-

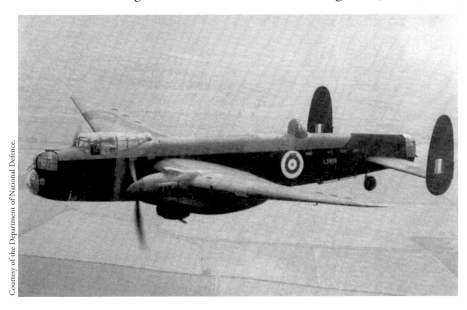

Courtesy of the Department of National Defence.

The Avro Manchester, predecessor to the Lancaster.

it with the Lancasters. But faced with American and British competition on the Atlantic, the final four bombers were redesigned as close to normal civil airliners as possible. The company tried hard to make them as comfortable as possible, and they were dubbed the "de-luxe" Lancs. Each had a "stand-up" cabin for the ten passengers in two rows, with yellow ceilings and walls, a mahogany-coloured carpet, fibreglass insulation, and combustion heaters. There were individual reading lights, call buttons, ashtrays, and oxygen outlets for each passenger. A galley was fitted to serve hot meals prepared by the first purser-stewards, or male flight attendants, at TCA, who, once out of Canadian airspace, introduced the airline's first ever bar service.

No longer restricted to the Dorval-Prestwick run, the Lancasters took on other cargo and routes, flying medicine to war-shattered Europe and war brides to Canada. On December 27, 1945, Canadian delegates chartered CF-CMY to travel to Havana, Cuba to participate in the first International Air

Ken Leigh Collection

Above: Aye, do ye ken Johnny Canuck? CF-CMS in camouflage landing at Prestwick with the first CGTAS mail. A defining moment in the company's (and country's) history.

Below: RCAF maritime reconnaissance Lancaster.

Courtesy of the Department of National Defence.

Transport Association conference held in the western hemisphere. To celebrate Aviation Day on July 12, 1946, as part of the city's Diamond Jubilee, Vancouver saw its first Lancaster; the flight from Prestwick took twenty-eight hours.

Avro was so impressed by the Lancaster's potential as an airliner that in December it bought Victory Aircraft Ltd. to build a civilian variant called the York. But only a single Avro York was built, since by then the North Star was being built at Canadair in Montreal. The beginning of the end for the CGTAS Lancasters came on November 16, 1946, when TCA took delivery of its first North Star and all the converted bombers were sold off overseas, ending their days as drudges for fly-by-night carriers. For many years, it was rumoured that two of the CGTAS Lancs were being used as gunrunners in the Middle East. But some evidence has come to light that the remainder may have taken part in the Berlin Airlift, carrying food and fuel to that beleaguered city. The RCAF brought its Lancasters back into service in 1952 as maritime reconnaissance aircraft, stationed at Comox, British Columbia and Greenwood, Nova Scotia, until Canadair Arguses replaced them in 1958. But whatever their fates, chief designer Roy Chadwick could never have imagined that of the thousands of Lancasters built, nine would serve as Canada's first Atlantic airliners — an honourable use of a great aircraft.

Bristol Beaufighter.

Courtesy of the Department of National Defence

BRISTOL BEAUFIGHTER

During the Munich crisis of 1938, there were many in Britain who recalled with fear the German Zeppelin and Gotha bomber raids of the Great War. The home defences then had been almost powerless to stop the air raids, especially at night, and the fear now was that, with the modern bombers of the contemporary Luftwaffe, the nation's cities would be laid waste with impunity. At the same time, the RAF was looking for a fighter aircraft that had the fuel capacity to escort its own bombers deep into Germany. The Bristol Aeroplane Company of Filton had just developed its dependable Hercules radial engines, which were to become its trademark well into the turboprop era. With them, it could produce the Blenheim bomber and adapt that design for a torpedo bomber. The Beaufort first flew on October 15, 1938, and, as it had the fuel capacity and armament for mine laying and anti-shipping strikes, Bristol designers realized that it could be adapted to the requirement for a night fighter and bomber escort.

Christened the Beaufighter, four prototypes were ordered just before the war began, the first making its initial flight on July 17, 1939. For a fighter, it was slow and heavy — the Bristol Hercules engines were then in short supply, and initially, inferior Merlin engines were used. But with four 20-millimetre cannon in the nose and eight .303-inch Browning machine guns in the wings, the Beaufighter was the most heavily armed fighter in the world. The RAF received its first Beaufighters in September 1940, in time for the German night attacks that winter. This was a totally new war, in which stealth in the darkness was more important than speed and manoeuvrability, and aircraft carried their own sensors with them rather than relying on searchlights and instruction from radar stations below. The Beaufighter made an ideal vehicle, both as a hunter and killer, and was responsible for ending the night air raids.

Canadian Beaufighter crew.

The bulky AI (airborne interception) radar sets that it carried required a second crew member, who would direct the pilot toward whatever signals they might pick up. Beaufighters were also built under licence in Australia to be used in the Pacific war. Instead of a pattern aircraft being flown out from Filton, 55,000 microfilm negatives were sent, and by the war's end, a total of 5,560 had been built.

Early in the war, the RCAF's three antishipping/antisubmarine squadrons — 404, 407 and 415 — were all woefully under equipped, with Blenheim and Hampden bombers. They were no match for the Luftwaffe Ju 88s they encountered, or the firepower from the flak ships that protected the enemy convoys operating off the Norwegian, Dutch, and French coasts. In September 1942, 404 Squadron, then at Dyce, Scotland, was the first RCAF squadron to replace its Blenheims with the IIF version of the Beaufighter. The long-range fighter bombers were equal to the Ju 88s and were pressed into use to disrupt Nazi communications before D-Day. Later versions were equipped with armour-piercing rockets, which, on October 9, 1944, 404 Squadron used to great effect against German merchantmen off the Norwegian coast.

Rocket-firing Beaufighters hit a German ship off Norway.

DOUGLAS BOSTON/HAVOC

Never to achieve the glory that its contemporaries like the B-25 and B-26 did, the Douglas Boston DB-7 carved out its own niche as a night fighter in the RAF "intruder" squadrons. The aircraft would never have been built at all if it wasn't for France and England, two of the many nations that used it. Others were South Africa, Australia, Brazil, Holland, and Canada, but none more so than the Soviet Union, which received 3,000 of the 7,478 made.

In 1936, Donald Douglas owned 51 percent of the Northrop Corporation and wanted to get into the light bomber market. He asked Jack Northrop and Ed Heinemann to design one, the whole project to be entirely funded by the company. They visualized the Model 7, an awkward-looking aircraft with a pair of Pratt & Whitney R-985 Wasp Junior air-cooled engines on shoulder-mounted wings. It was the first American military aircraft to have a tricycle undercarriage. Depending on the customer's requirements, it could have either a glazed, lowered midsection to be used as an observation platform or a bomb bay for up to one thousand pounds. Its limitations were that it was underpowered and had a restricted range, and in late 1936, Douglas put the project away to concentrate on the B-18 bomber instead. A year later, the United States Army Air Corps asked for an attack bomber, specifying that with war imminent, it had to be built immediately. The British and French were also shopping around for light bombers. Heinemann dusted off the Model 7 project and added a few refinements: more powerful Pratt & Whitney R-1830 engines and a choice of two nose sections (transparent for bomb aiming or packed with machine guns in the attack version). But by the time the 7A flew on October 26, 1938, the United States Army Air Corps was enamoured with

Douglas Boston/Havoc.

the B-25 and showed no interest. Fortunately, the French Purchasing Commission was also at the demonstration, and they ordered 270 DB-7As. The British Purchasing Commission, too, liked what they saw, and on February 20, 1940, placed an order for 150 aircraft, calling theirs Bostons. They specified replacing the Pratt & Whitney engines with Wright R-2600 Double Cyclones. These gave the Boston the power it needed: 1,600 hp at takeoff and 338 miles per hour at 12,000 feet. Designated Boston III or Havocs, they replaced the Blenheims in the RAF medium bomber squadrons and were bombing targets in France by February 1942.

Bostons that were sent to the Intruder or Turbinlite night fighter squadrons were painted in matte black, their engines were fitted with flame-dampeners, and they were armed with a belly-pack of four 20-mm Hispano cannons. They ranged over enemy airfields, ambushing Luftwaffe night fighters as they returned to base. Canada acquired only three Bostons, one each of the III, IIIA, and IV models, but the RCAF 418 Squadron used Boston IIIs from the British stocks. Based at Debden, Essex, it was this squadron, using Bostons and Mosquitos, that ranged deep into enemy territory, achieving one of the highest scores in aircraft and flying bombs destroyed.

Grumman Avenger.

GRUMMAN AVENGER

Named the Avenger because the United States Navy ordered Grumman to build the aircraft immediately after the attack on Pearl Harbor, the torpedo bomber would become the navy's mainstay after the Battle of Midway. By 1937, the company had expanded to the fabled "Iron Works" at Bethpage, Long Island, and Grumman was turning out F4F3 Wildcats: rugged, slow, carrier-borne fighters. When the United States Navy wanted to replace its standard torpedo bomber, the TBD-1 Devastator, Grumman adapted the Wildcat design into a candidate.

The Avenger TBF was another generation removed from the Devastator, and much of the credit was due to the design and engineering team led by W.T. Schwendler. It might have been slow and sluggish, but, unlike the Devastator, it had rearward folding wings that allowed it to fit into the holds of small carriers. Yet it was large enough to carry a crew of three (pilot, observer, and rear gunner) and deliver a twenty-one-inch torpedo or two thousand pounds of bombs. Pilots thought the Avenger was pleasant to fly and used it in a variety of roles — antisubmarine, dive-bomber, search and rescue, and, much later, an electronic countermeasures platform. Following the Battle of Midway, the Avenger replaced the Devastator, and its range and speed were used to good effect by Admiral Marc A. Mitscher's TF58 at the Battle of the Philippine Sea. The Avenger was credited with the sinking of forty-five U-Boats and two Japanese battleships. Called Tarpons (after an armoured Caribbean fish) by the British, who received lend-lease Avengers in 1943, they served in Royal Navy until 1962. Leroy Grumman, who valued the family relationship with his employees, kept his company small through the war, building only 2,293 Avengers at Bethpage. It was Eastern Aircraft, part of General Motors, that turned out the majority of the planes (7,482).

Other countries that received Avengers during and after the war were New Zealand, the Netherlands, France, Canada, and even Japan. Canadian Avengers flew from the aircraft carrier HMCS *Magnificent*, and by 1957, they were assigned to the naval reserve air group. Beginning in 1958, their survivors were sold off to be adapted as water bombers and crop-dusters, used especially in New Brunswick.

LOCKHEED CONSTELLATION

Since the birth of commercial aviation in the United States, every airline and every aircraft manufacturer has understood that the main route was the transcontinental one, connecting the two coasts of the country. Using Ford Trimotors and Curtiss Condors, the major airlines raced to fly from New York to Los Angeles, Chicago to San Francisco, or Boston to Seattle, with as few refuelling stops as possible. The average times to complete the twenty-six hundred miles from coast to coast varied, from an American Airlines DC-3 in 1937 (sixteen to twenty hours) to a Trans World Airlines Boeing 307 Stratoliner in 1940 (fourteen hours).

As Douglas had captured the twin engine market with its DC-3, in 1939, Lockheed began work on a four-engined airliner that would not only be pressurized but have the range to fly from New York to Los Angeles without a single stop. Code named "Excalibur," the project attracted Howard Hughes, the principal stockholder in Trans World Airlines (TWA), who would collaborate with Lockheed's Hall Hibbard and Clarence "Kelly" Johnson in its design. The story that the eccentric Hughes absentmindedly drew the aircraft's sensuous S-shaped fuselage on the back of an envelope while he was directing Jane Russell in the movie *The Outlaw* in 1941, is just that — a story. While the Constellation owes part of its birth to Johnson's previous masterpiece, the P-38 Lightning twin-boomed fighter, it was a young Lockheed engineer, Ward Beman, who actually shaped the aircraft's fuselage to correspond to the airflow along its body. The 049's trademark would be the swept-down, protracted nose that flowed upward into a slender fuselage and stretched for ninety-five feet, ultimately spiralling into a triple tail. With the war, the American military placed an order for 313 aircraft, which, in the Lockheed tradition of naming their planes after stars, were

Lockheed Constellation.

called Constellations, to be shortened to "Connies." The first flight took place on January 9, 1943, and by the war's end, both Pan American and TWA were flying the civil version, or 649, either across the Atlantic or across the United States (on which route they were three hours faster than the DC-4s of American Airlines).

Committed to the Canadair North Star after the war, Trans Canada Airlines knew of the Constellation only through its use by BOAC, Air France, and KLM on their Montreal runs. When TCA looked to replace its North Stars in 1951, it considered both the Douglas DC-6B and the Lockheed Constellation, now in the 1049 series. Airline personnel were hesitant, remembering the problems with previous Lockheed aircraft like the Electra and Lodestar, and the Wright Cyclone R-3350 engines that the 1049 used were known to be troublesome and difficult to maintain. Nevertheless, after due consideration, fourteen Constellations were bought by TCA: five 1049C (CF-TGA to CF-TGE), three 1049E (CF-TGF to CF-TGH), four 1049G, with their wingtip fuel tanks (CF-TEU to CF-TEX), and two 1049H (CF-TEY, CF-TEZ). They served from August 1954 to January 1962; ironically, they would be displaced by another Douglas aircraft that TCA ordered, with Rolls Royce engines — the DC-8.

Wartime Constellations lined up at Lockheed in Burbank.

Who speaks for Canada? In the late 1940s, there were few sights more Canadian than this TCA North Star and its crew.

CANADAIR NORTH STAR

As Canadian as the *Bluenose* schooner or the RCMP Musical Ride, the Canadair North Star was Canada's own airliner. It was to the post-war Dominion what the birch bark canoe had been to colonial British North America. The C-54 airframe it used was already dated when Canadair assembled the first North Star, and its four Merlin engines were better suited to a bomber. But if ever an aircraft came to define a country and symbolize its hopes for the future, it was the North Star.

Faced with Hobson's choice (keeping the civilianized Lancasters in service until there were Douglas DC-6 airliners available or buying the inferior British Avro Tudors), the post-war Canadian government opted to build its own aircraft. British and American aircraft technology had been blended together before and produced the North American Mustang aircraft, perhaps the best fighter aircraft of the Second World War. Perhaps it was the success of that marriage that was the impetus for attaching four Rolls Royce Merlin engines to a Douglas DC-4 airframe. The idea was a sound one: the DC-4 was already in use with the military and a dozen airlines then plying the Atlantic. The Merlin engines came from the Rolls Royce stables, a British company with an impeccable pedigree. And as C.D. Howe, the minister for reconstruction, explained to the House of Commons, they had been chosen to help the impoverished Mother Country in its time of financial need. Eager to break into the North American market, Rolls Royce even agreed to a "never be sorry" clause: it would repair whatever went wrong with the Merlin without charge.

The C-54 airframes were sold to the Canadians at the garage sale prices of two hundred dollars a ton and freighted to Montreal from the Douglas factory at Parkridge, Illinois. The Douglas company stipulated that whatever aircraft was

Courtesy of the National Archives of Canada C85168

C.D. Howe at the official launch of the North Star, July 18, 1946. Man and machine share the same characteristics — noisy, rugged, dependable, almost obsolete.

built out of the parts would be used domestically or on international routes from Canada and not in competition with Douglas's own aircraft. The airframe and engines were assembled by the Canadair company in the Montreal suburb of Cartierville, and the crossbreed was christened North Star by Alice Howe, the minister's wife. The North Star was very much C.D. Howe's pet project and part of his planned economy. The RCAF was ordered to take twenty-four North Stars in the unpressurized version, and Trans Canada Airlines would take twenty, which were to be pressurized. Howe hoped that there would be enough inducements to get Canadian Pacific Airlines to take some as well. It was an imaginative solution to the country's needs and, in 1945, about all that its industrial base was capable of.

After the Lancaster bombers, TCA embraced the North Star as its own. Noisy it may have been, but it did seat thirty-six passengers, giving them, as the TCA brochure said, "plenty of room to walk about in." It was the first four-engined aircraft ever built in Canada expressly for the purpose of carrying passengers. For the only time in its (or Air Canada's) history, the airline bestowed the names of heroes from Canadian history on its aircraft: Jacques Cartier, Samuel de Champlain, Edward Cornwallis, and Alexander Selkirk. Even the infamous Merlins, the powerplants that deafened so many passengers, received some immortality. The poet Earle Birney compared them to the steeds of the Greek figure Bellerophon, who rode the winged horse Pegasus.

Assembling the North Star did keep Canadair functioning until the military contracts for F-86 Sabres came through during the Korean War. Instead of just forty-two North Stars, seventy-one were built between July 1946 and July 1950. Twenty went to TCA and twenty-five to the RCAF (the last being the C-5, the North Star with Pratt & Whitney engines), four were sold to Canadian Pacific Airlines, and twenty-two were exported to the British Overseas Airlines Corporation, where they were dubbed Argonauts. By taking over the first six North Stars from the RCAF order, TCA had the first one in service on November 16, 1946, receiving its last in June 1948. With this fleet of "Skyliners," it was able to expand to Paris, Bermuda, and the Caribbean. For the first time, Canadian ministers could visit other countries in Canadian-built aircraft (as Lester Pearson did) or travel around the world (as Prime Minister John Diefenbaker did in December 1958). When a North Star flew the first Canadian-born governor general, Vincent Massey, on a tour of the Canadian Arctic and over the North Pole on March 24, 1956, the symbolism was obvious. Canadian Pacific Airlines used their Canadair Fours from 1949 to 1951 to pioneer the routes to China, Hong Kong, Fiji, and Australia. As soon as the RCAF received its North Stars, 426 Squadron used them to airlift troops to the Korean War. The North Stars later carried Canadian troops on United Nations peacekeeping missions to the former Belgian Congo, Cyprus, and the Middle East. The RCAF gave up its last North Star in 1966, replacing them with Canadair Yukons.

Trans Canada Airlines began phasing their North Stars out in 1960, parking them at Dorval Airport, where they had first flown in TCA colours fourteen years before. In the West Indies, North Stars were phased out on January 28, 1961. The final North Star to carry TCA passengers touched down in Montreal on April 30, 1961, and the last North Star flight carrying cargo ended, appropriately enough, on Dominion Day in 1961. The entire fleet of TCA North Stars sat forlornly at Dorval, until eleven were bought by the British aircraft brokers Overseas Aviation, and five by Mexican freight carriers Lineas Aereas Unidas. Of those sold to Overseas Aviation, one played a part in the shadowy world of gunrunning. In February 1965, on the suspicion that a rogue CIA agent was using it for smuggling weapons into Algeria, the Dutch authorities impounded a Panamanian-registered North Star at Schipol Airport in Amsterdam. Through the spring rains, the Dutch watched successive coats of paint wash off the aircraft until its original TCA markings reappeared. The owners paid the fine and left with the North Star, but in October of that year, the old aircraft reappeared in the Netherlands, this time at Rotterdam Airport, to pick up more guns. According to the flight plan, it was going to Birmingham, so the Dutch allowed it to take off. The North Star, three thousand pounds overweight, struggled into the air and made for Majorca instead. There it refuelled for a flight to Fort Lamy, Chad — this time the guns were bound for the breakaway Nigerian state of Biafra. On the way, it ran out of fuel and crashed near Garona, Cameroon. The former TCA airliner broke up into four sections, scattering arms and ammunition over the swampy terrain. Its crew suffered mild concussions, testimony to a well-built aircraft. There is a single North Star — an RCAF C-54M at the National Aviation Museum in Ottawa — in existence today: a sad end to an aircraft that has come the closest in Canadian history to being a national symbol.

Courtesy of the Department of National Defence.

RCAF North Star.

Bristol Freighter.

Courtesy of the Department of National Defence

BRISTOL FREIGHTER

When railway and tram car manufacturer Sir George White decided, in a fit of civic pride, to rename his British & Colonial Aeroplane Co. after his city of Bristol, no one could have expected that it would amount to very much. It was 1910, and airplanes were an unknown means of transport. The first aircraft the company built was the Bristol Boxkite, a pirated copy of Henri Farman's plane, but in the succeeding years it produced a series of aircraft like Scouts, Bulldogs, Blenheims, and Beaufighters. The company trademark, it seemed, was to manufacture ugly, resolute, powerful aircraft, and thus homeliness of its Freighter came as no surprise to anyone.

The Bristol 170 Freighter, designed in World War II, arrived too late for the conflict. Its cavernous maw and clam-shell nose doors had evolved from Bristol's 130 Bombay troop transport of 1935, and like its predecessor, the Freighter had the potential to be a transporter of not only military vehicles but civilian ones as well. The aircraft's two Bristol Hercules sleeve-valve engines could lift up to three saloon cars or twelve thousand pounds of freight at a minimum speed and cost; the British airline Silver City Airways used its Freighters as cross-Channel air ferries. Its hold was ideally suited for this — in a design that was to be later followed by the Boeing 747, the Freighter's crew sat above the cargo. The cockpit could only be reached by a ladder that led up through a hatch behind the first officer's seat. Despite its curmudgeonly outward appearance, the Bristol 170 was considered pleasant to fly.

To meet the need for a dedicated freighter in 1953, Trans Canada Airlines bought the three Bristol 170s (CF-TFX, TFY, and TFZ), which allowed for an almost daily service on the Montreal-Toronto-Lakehead-Winnipeg and Montreal–New York–Toronto routes. TCA's engineering department extensively modified their Freighters, adding a

Canadian electrical system, better heating, ice protector plates on the fuselage sides, restraining gates in the cargo hold, a loading winch, and flood lights to allow for checking the ice on the wings at night. Regarded by American air traffic controllers with some amusement, the TCA Freighters nevertheless fulfilled every expectation until replaced by converted North Stars. The air freight market that they had been purchased for never materialized, and they were sold off to bush carriers. Militarily, Bristol had greater success with its 170.

The Pakistani, Argentinean, Australian, and New Zealand air forces operated Bristol Freighters, as did the RCAF from 1952 to 1967. Its fleet of six Freighters shuttled between RCAF bases at home and in Europe with personnel, engines, and disassembled aircraft. In 1968, the last survivor of the six RCAF Freighters was sold to Max Ward, and in his service it became the first Canadian aircraft to land at the North Pole. The Western Canada Aviation Museum has a Bristol Freighter on display, the former RCAF 9699.

Bristol Freighter in TCA livery.

GRUMMAN ALBATROSS

Although it was generally acknowledged that the day of the flying boats had ended with the Second World War, many air forces continued to operate Sunderland, Marlin, and Convair flying boats well into the 1960s. While they had been superseded in antisubmarine warfare by landplanes, smaller flying boats were still necessary for in-shore search and rescue, and by 1957, the RCAF was actively looking for a replacement for its venerable Cansos in this role.

A decade earlier, the United States Air Force had put out a requirement for an amphibious aircraft that could operate equally from land, sea, or, with skis, snow. With its Goose series, the Grumman Corporation had experience in the field and developed the Albatross from that, the prototype flying on October 24, 1947. The United States Air Force, Navy, and Coast Guard ordered a total of three hundred aircraft, designated the SA-16. The aircraft were used to good effect during the Korean War to pick up and rescue personnel from rivers and coastal waters. Powered by two Wright R-1820-82 radial engines, the Albatross could carry 8 to 12 passengers, cruise at 165 miles per hour, and travel for 1,650 miles — longer with drop fuel tanks. JATO and reversible propellers assisted taking off and manoeuvring on rough seas. The RCAF opted to purchase the Albatross designated as the CSR-110 in 1960, using it on both coasts for a decade, until Buffalos replaced it. The Albatross was the last amphibian to be used by the RCAF; none remain in Canada today, as other countries were quick to purchase them when they were declared surplus. In 1979, there were still an estimated two hundred Albatrosses flying, and Grumman offered a conversion package to reconfigure them as twenty-eight-seat commuter passenger aircraft for resorts, and later as water bombers powered by Garrett turboprop engines.

Grumman Albatross.

CANADAIR SABRE

The eye-pleasing F-86 Sabre was what the public thought a jet fighter should look like. The undisputed champion of the free world during the Cold War, it had as much to do with Nazi research into aerodynamics as it did with the United States Navy wanting a jet fighter in 1944. After the war, North American Aviation was still riding the popularity of its P-51 Mustang and was eager to break into the jet era. Its first jet aircraft, the FJ-1 Fury, built for the United States Navy, did get to Mach .87 — the highest speed ever for an American "straight-wing" fighter — but the company understood that it would never go above six hundred miles per hour. Then captured German data showed that a swept wing would delay the compressibility effects that an aircraft experienced at high subsonic speeds. Wind tunnel tests allowed North American's engineers to decide on a wing and tail that swept back at an angle of thirty-five degrees. The redesigned Fury, now designated XP-86, flew on October 1, 1947, later exceeding Mach 1 (the speed of sound) in a dive. The swept wings were kept as "clean" as possible with the armament of six .5-inch machine guns mounted on the sides of the nose. As with all early jets, the Sabre had little internal fuel, so to increase its endurance, wing drop tanks were added.

When the United States Air Force ordered it, the designation was changed to the F-86A Sabre, and in December 1950, the first units were rushed to Korea to stem the Communist advance in that country. There, the swept-wing MiG-15s showed that the Soviets too had benefited from German technology. On December 17, 1950, the inevitable occurred: the first known combat between jet fighters took place as Lieutenant Colonel. Bruce H. Hinton shot down a MiG-15. The Sabre came to represent the United Nations' success in the Korean War, as it was the

Canadair Sabre.

only transonic swept-wing aircraft to match the Communist MiG. It brought a whole set of innovations like pressurization, screen de-icing, g-suits, and a radar on the upper lip of the nose intake. Pilots thought it an unforgiving aircraft, which, because of its wing leading edge slats, had to be carefully flown. In the early models, the nose had to be held up at an angle of attack of at least thirteen degrees, or the aircraft would sink. It did reach supersonic speeds in a dive, but if the dive continued below twenty-five thousand feet, it had a tendency to roll, a problem that was difficult to check, owing to aileron reversal.

Canadair plant with freshly minted Sabres on tarmac.

Courtesy of the Canadair Collection

But what the Supermarine Spitfire had been to Britain during the war, the F-86 Sabre came to be to Canada in the 1950s. Until it started production of the Sabre in August 1949, Canadair had only assembled DC-4 airframes and refurbished DC-3s. To take on an aircraft that embodied the most advanced designs, systems, and materials of the time was a considerable technical feat, which was all the more remarkable when one considers that much of the Sabre was still on the United States' "Top Secret" list. Despite this, the first Canadian Sabre F-86A was completed in July 1950 and used by test pilot Al Lilly to break the sound barrier on August 10, 1950 (the first Canadian to do so). Of the twenty-two RCAF pilots who were attached to the United States Air Force to fly Sabres in Korea, some, like F/L J.A.O. Levesque, shot down MiG-15s. When, in September 1952, RCAF squadrons in Canada received the jet fighter, among the first were 421 and 410 Squadrons at St. Hubert, the latter packing their Sabres onboard the aircraft carrier HMCS *Magnificent* for shipment to Europe soon after. The year 1952 was unmistakably the year of the Canadian Sabre. On May 30, in operation "Leapfrog," 439 Squadron flew twenty-one Sabres from Uplands, Ottawa in stages to North Luffenham, Britain. By June 1952, Canadian Sabre 2s were flying in Korea with the United States Air Force and were said to be preferred by the American pilots to their own. As No. 1 Air Division was formed in Paris that summer, the RCAF arrived in strength in Europe and No. 2 Fighter Wing (416, 421, and 430 Squadrons) flew from Canada to their new home at Grostenquin in Germany. In Europe, the Canadian pilots discovered that, for the first and only time, with their F-86s they were in the vanguard of technology. The French could only field their slow-moving Ouragan jets, and the British Supermarine Swifts had proven a disappointment. For a brief and gratifying period, Canadian Sabres ruled the NATO skies. Four hundred Sabre 4s were supplied by Canadair to the RAF, and later, as the RCAF received its Sabre 5s, its older Sabre 2s were given to Greece and Turkey. What most Canadians recall of the beautiful Sabre was its use in the RCAF's aerobatic team, the "Golden Hawks," which flew them in air shows from 1959 to 1963. The shark-nosed fighter was to be the most successful Canadian aviation export program ever. Canadair Sabres were sold to the Germans, the South Africans, and the Colombians. Years later, these same aircraft would be flying with still other air forces: Yugoslavia, Italy, Pakistan, and the Honduras. Ironically, during the India-Pakistan conflict, the Canadian-built Sabres flown by Pakistani pilots would be pitted against Indian Air Force British-built Hawker Hunters and would consistently outmanoeuvre them — something that their manufacturer could never have foreseen.

The last Canadian Armed Forces Sabre flew in 1969, and its storage ended the most glorious era in Canadian aviation history.

CANADAIR CT-133 SILVER STAR

More than likely, every Canadian Armed Forces base and aviation museum in Canada has a Silver Star somewhere on the premises, either on display within the museum or at the aero park or "flying" from a pylon. Eclipsed by its contemporary, the more eye-catching Sabre, the Silver Star trainer was the quiet workhorse of the RCAF, outlasting the Sabre in air force service by more than three decades.

In 1947, jet powered fighters were just out of the realm of science fiction. Only the British operated them, and when Lockheed flew its first jet fighter, the P-80 Shooting Star, the number of crashes was alarming. The problem was that, with the exception of the two-seat Gloster Meteor, there were no jet trainers. To cut down on the crashes, Lockheed built a forty-six-hundred-pound thrust, Allison-engined, two-seat trainer from the P-80, fitting an instructor into the expanded aircraft. This meant reducing the size of the fuselage fuel tank, and, to compensate, wing tip tanks were added. Ironically, with these modifications, the two-seat trainer designated as the T-33 was so good that its performance was better than the fighter. The T-33 would be the United States Air Force's only jet trainer for a full decade.

The first T-33s to come to Canada were thirty trainers borrowed from the United States Air Force in 1951, the same year that Canadair was granted a licence to build the aircraft. The RCAF issued an order for 576 T-33s on August 13, 1951, later increasing this to 656. In place of the Allison J33-35 turbo-jet, Canadair used a fifty-one-hundred-pound thrust Rolls Royce Nene 10 built by Orenda. The Canadian version, designated T-33 Silver Star (later CT-133) was used for pilot and armament training and photo reconnaissance. On December 22, 1952, W.S.

Canadair CT-133 Silver Star.

Longhurst test-flew the T-33 prototype at Cartierville, the first of 656 "T-Birds," as they were called, to serve with the RCAF and CAF until 2000. They were exported to France, Greece, Portugal, Turkey, and Bolivia, where they operated against guerillas. The most famous T-Birds were part of the "Red Knight" aerobatic team of the 1960s. Silver Stars also served with the RCAF in NATO, arriving in 1953 at Baden, Grostenquin, and Zweibrucken. While primarily an advanced trainer for Sabres, CF-100s, and later CF-104s, the Silver Star was also a utility aircraft, instrument trainer, and Mobile Repair Party support. The arrival of the CF-18, a dedicated dual seat aircraft with its flight simulator, meant that the old jet trainer could now be retired, but there was one final hurrah — pilot training during the Gulf War.

The last CT-133 flight at Baden-Soellingen took place on March 31, 1992, before the base closed, and by April 8, all had flown home to Canada. The advent of electronic warfare gave the CT-133 a new lease on life, in the aggressor role or as a target for CF-18s. It was posted to 4 Wing, Cold Lake Alberta; 3 Wing, Bagotville, Quebec; 19 Wing, Comox, British Columbia; and 14 Wing, Greenwood, Nova Scotia. At Cold Lake, for example, 417 Combat Support Squadron used its eight CT-133s as adversaries for the CF-18 and in Close Air Support Training. In 1998, the Silver Star, the most successful jet trainer in history, celebrated its fiftieth anniversary in the Canadian forces. They were phased out on March 31, 2002.

Clarence "Kelly" Johnson, who designed many of the aircraft in this book, with a model of his T-33.

Sikorsky Helicopter.

SIKORSKY S-55

The first use of a helicopter in Canada took place on May 19, 1945, when an American military Sikorsky R-4 was airlifted from New York to Labrador and used on a mercy mission. The RCAF bought Sikorsky S-5s soon after, but it was the Royal Canadian Navy that equipped itself with the S-55 model. In the 1950s, the RCN operated two aircraft carriers: the HMCS *Magnificent* and the HMCS *Bonaventure*. To provide plane guards on each (i.e. pick up the pilots who had ditched on landing or takeoff), in 1951 it ordered thirteen S-55s, the naval version of the military H-19 Chickasaw helicopter. The H-19 was the best utility helicopter then available and was being manufactured under licence by Westland in Britain as the Whirlwind, the machines still in use forty years later in the British Army as the Wessex. The S-55 was unique for its time in that its Wright R-1300-3 engine was held in its bulbous nose behind clamshell doors, leaving the body available for nine passengers. The RCN S-55s arrived at CFB Shearwater in July 1952, before being sent to the carriers. Earning the nickname "Angel" for picking up ditched pilots, they lifted off the flight decks of both aircraft carriers until decommissioned in 1966.

The RCAF ordered fifteen S-55s for search and rescue and to build the now almost forgotten Mid-Canada Radar Line. Also known as the McGill Fence, the Mid-Canada Line was made up of ninety-eight Doppler radar sites in musk keg and mountain, between the DEW and Pinetree Lines. The sites were cleared by tractor trailers, helicopters, and landing craft, but the Piasecki and H34 helicopters that the RCAF had were insufficient and undependable, so the SAR S-55s were employed. Okanagan Helicopters, familiar with their use at the Kitimat project, was contracted to train more S-55 pilots. In their distinctive blue and red colour scheme, the S-55s serviced the McGill Fence until its closure in 1965. Replaced in the Canadian Navy in 1963 by the Labrador and the Sea King, many S-55s remain in use commercially, converted to turbine power for a further lease on life.

GRUMMAN TRACKER

I f any aircraft typified the Grumman Iron Works, the Tracker did. Built to operate from aircraft carriers, in Canadian service the Grumman Tracker had the misfortune to live in interesting, ever-changing times. It would spend most of its operational life flying from shore, adapting to shifting political priorities, and flying in roles that it was never designed for.

In 1950, to protect its carrier battle groups, the United States Navy tendered for an antisubmarine aircraft able to operate from a small carrier deck, powerful enough to carry the necessary detection equipment, yet small enough to fit below. With its experience of carrier aircraft, Grumman obliged with the S-2A Tracker, a compact airframe with folding wings powered by two hulking 1,475 hp Wright R-1820 nine-cylinder air-cooled engines. The Royal Canadian Navy was then considering a replacement for another Grumman aircraft, the Avenger, to operate from its only aircraft carrier, the HMCS *Bonaventure*. The Tracker was chosen, and Ottawa specified that one hundred aircraft designated CS2F-1/2 be made under licence by De Havilland Canada in Downsview, Ontario. The first was handed over to the Royal Canadian Navy on October 13, 1956, for use on the *Bonaventure*. When Canada's last aircraft carrier was retired in 1969, it looked like the end for the Tracker as well, but such was the aircraft's popularity that the newly unified Canadian Armed Forces raised enough of a fuss to keep it on. A good platform for any equipment needed and strong enough to fly low, it could be adapted for in-shore coastal patrol, complementing the CP-140. The enemy to Canadian shores was no longer Soviet submarines but ships bringing illegal immigrants, fishing poachers, smugglers, and potential environmental disasters. With arresting hooks removed, the Tracker, now the CP-

Tracker landing on HMCS Bonaventure *EKS 1487.*

Courtesy of the Department of National Defence

121, was conditioned to operate from shore bases like CFB Comox and Summerside. The antisubmarine equipment was removed, and a distinctive black radome housing search radar was added. Then the old aircraft were given new teeth in the form of Bristol Aerospace CRV-7 air-to-surface rockets for use in their antishipping role.

In 1988, when it looked once more as if the Tracker's days were numbered, IMP Aerospace of Halifax was contracted to re-engine them with Pratt & Whitney turboprop engines. The company had previously received an order to upgrade twelve S-2E Trackers for the Brazilian Air Force. Only one CF Tracker was converted before the project was cancelled, and in 1990, the Trackers were finally, irrevocably struck off. They ended their days in Canada and the United States, with Conair adapting several as Firecat water bombers — following yet again the Grumman Avenger, which they had displaced in naval service. In the United States, the Marsh Aviation Company in Arizona re-engined former United States Navy Trackers with Garrett turboprops, selling them as water bombers and maritime patrol, the latter to the Armada Argentine in 1993.

Avro Arrow.

AVRO ARROW

"They built us the Arrow, they promised us the Sparrow,
They promised us SAGE and the ASTRA as well,
Along came old Georgie, along came young Johnnie,
And Arrow and Sparrow were shot all to HELL."

409 Squadron Ode

Other countries have overreached themselves with aeronautical creations, notably the British with the BAC TSR 2, and the Americans and the Russians with their SSTs. But none have made a cult out of it, as Canada has. Forty years later, it remains impossible to write of the Avro Arrow without stirring up the emotions of some Canadian somewhere. Whether one regards the CF-105 Arrow as a technological marvel twenty years ahead of its time or whether one puts it in the same category as the RMS *Titanic* (i.e. excessive national ambition that sank on its maiden voyage), no one seems to be neutral.

Avro's success at developing both the Canuck and the Jetliner in record time encouraged the RCAF in April 1953 to give them an impossibility. To meet a perceived threat of Soviet bombers, the air force issued AIR-73, a requirement for an all-missile, twin-engined, two-seated supersonic interceptor with a complex fire control and radar system, an action of radius of 620 miles, and a speed above Mach 1.5. That no one, not the Americans, the British, nor the Russians, had developed such an aircraft did not seem to faze the RCAF. It wanted six hundred interceptors within an allotted time span and would accept nothing less. To the eternal credit of Avro's talented engineers, they pulled it off. So much of what was required by AIR-73 was unknown that what this band of brothers accomplished has become the stuff of legend: the thin delta wing, the rudimentary "fly-by-wire" control system, the use of titanium, the Orenda Iroquois engines ... all showed what Canadians could do. What tripped the project up were things that Avro had no control over: the Astra radar system built by RCA-Victor, the Sparrow II missile by Douglas-Bendix, an unsympathetic Toronto press — and a fatal change of government in Ottawa.

When the first Arrow was rolled out at 2:00 P.M. on October 4, 1957, Minister of Defence George Pearkes could rightly claim that a new age was upon them. Little did he know how prescient his words were. That May, the Conservatives had come to power, and Pearkes did not have the knowledge or the faith that his predecessor Brooke Claxton had had in the Arrow. Worse, that very day the Soviets had put their first Sputnik in orbit. In the shock and confusion of the Soviet Union's lead in missile technology, manned bombers were thought (wrongly) to be as obsolete as dreadnaught battleships, and equally wasteful. If they were unnecessary, then so was the development of interceptors like the Arrow, the cost of which had jumped per plane from $2 million to over $12 million. The Canadian military had always harboured resentment toward Avro for being "out of control" and had long felt that the funds allocated could be better spent on more conventional weaponry. The Arrow could not take the SAGE computer system and had no potential for export to either the British or the Americans, both of whom would soon have their own supersonic fighters to sell. On August 25, 1958, the military — citing the cost overruns of two billion dollars over the next three years, at a time when the whole defence budget was only six billion dollars, and the logic of matching Soviet missiles with an anti-missile system — recommended to the prime minister that the Arrow program be cancelled. Newly elected John Diefenbaker, aware of the fifteen thousand unemployed people that this would leave in the Toronto ridings, bought time by trimming it: the radar and fire control systems were cut on September 28, 1958. It wasn't as though Canada would be defenceless, as the Americans offered Ottawa the Bomarc ground-to-air missile instead. On February 3, 1959, with the next fiscal year on the horizon, Diefenbaker told the Cabinet that the decision could no longer be delayed. On February 20, at 11:00 A.M. on "Black Friday," he announced that the CF-105 Avro Arrow was to be terminated.

Perhaps, in a manner similar to a well-known Agatha Christie story, everyone — the politicians, the press, and the military — contributed to the death of the Arrow. Public opinion has laid the blame squarely on Prime Minister John Diefenbaker, who, while he campaigned on independence from the Americans, quailed when it came to paying for it with the Arrow. Other villains cited are the uninformed George Pearkes, the inflexible RCAF, the chairman of the chiefs of staff General Charles Foulkes, and finally those favourite whipping boys of Canadian history — the Americans, who can always be counted on to have an agenda of their own.

AVRO ARROW

By its untimely demise, the Avro Arrow has now, like the *Titanic*, spawned not only an industry devoted to keeping the memory alive, through the merchandising of T-shirts and mouse pads, but also endless speculation about "what might have been." Was it Canada's Great Betrayal or its last chance at breeding its own fighter aircraft industry — as the Swedes have? The argument continues ...

Canadair CL-41 Tutor.

CANADAIR CL-41 TUTOR

In the mid 1950s, aircraft manufacturers in several countries were designing dedicated jet trainers. All anticipated a huge market as air forces looked for replacements for their post-war T-33s and Meteors. In the United States, the candidates were the Cessna T-37 and the Rockwell Buckeye; in Britain, the Hunting Jet Provost; in Italy, the Aermacchi M.B. 339; in Yugoslavia, the Soko Galeb; and in France, the Fouga Magister. All offered a basic jet training aircraft that could also be used as a light attack fighter. Realizing that the RCAF would soon be considering a jet trainer, in 1956, Canadair began work on the CL-41 Tutor as a private venture. The RCAF was considering the imaginative scheme of eliminating piston-engined aircraft for primary training altogether, with the pilot using only one jet trainer.

Initially, the Pratt & Whitney JT12 engine was chosen to power the Tutor, but at the insistence of the federal government, the General Electric J85-CAN-40 was substituted. Heavier than the JT12, the J85's saving grace was that it was built by the Canadian Orenda Company. The first Tutor CF-LTW-X flew on January 13, 1960, and the RCAF ordered 190 aircraft, with deliveries beginning on December 16, 1963. Then the air force discovered that it could not do without basic piston-engined trainers, and in 1967, it ordered Beech Musketeers, keeping the Tutor for advanced jet training. Hoping to keep the line open by attracting other customers, Canadair launched an aggressive export campaign. It marketed the aircraft around the world in various guises: as a shipboard fighter, a counter-insurgency aircraft, and, as the airlines were entering the jet age, a commercial pilot trainer. Successful only in the Netherlands and Malaysia (where it was called the Tebuan or Wasp), Canadair's salesmen could not compete with the foreign manufacturers. The Canadian Armed Forces received the last CL-41 in September 1966.

Now redesignated the CT-114, forty years later the Tutor is still in use as a basic trainer. But it is more familiar to the public as the mount of the CAF Snowbirds aerobatic team, with the famous red and white colours and smoke generators. The Tutor is long overdue for replacement, and the British-built Hawk is seen as the likely candidate.

Tutor Snowbirds in formation.

Courtesy of the Department of National Defence

CC-130 HERCULES

Named for the Greek mythological figure who was strong and versatile enough to do anything, the Hercules has become synonymous with airlift. As analogous with the United States as Coca Cola and Hollywood, the Hercules is sought by American friends and foes equally, in over sixty air forces worldwide. "Anytime, anywhere" could be the unofficial motto of the "Herk" squadrons. First flown on August 23, 1954, the Lockheed Hercules remains in production fifty years later, a record unequalled by any other aircraft. What the C-47 was to the Second World War, the C-130 has been to modern wars and disasters, performing in civil and military roles as diverse as forest service, polar supply, weather reconnaissance, satellite recovery, and crop spraying. The concept of the Hercules was born during the Korean War, when the C-119 Flying Boxcar was the mainstay heavylift aircraft of the United States. There was some opposition in Lockheed to building a straight-wing ugly duckling of a freighter that still had propellers, sat squarely on its fuselage, and had a "Roman nose" — in short, that had none of the modernity of the swept-wing jet transports about to come into service. The production line was set up in a Marietta, Georgia plant, freeing up the main facility at Burbank, California to concentrate on more profitable enterprises. Like the C-119, the C-130A was also high wing and had aft loading, but its three-bladed Aerotech props were known to be unreliable. When the United States Air Force tactical squadrons traded in their Flying Boxcars for the new aircraft in 1956, they discovered to their delight that the four Allison turboprops allowed them to take off from dirt strips in only eight hundred feet. The tandem main landing gear with the low pressure tires in pods on the fuselage sides allowed for an unobstructed interior. The 130 was an instant success, and the B model was equipped with the four-

CC-130 Hercules.

bladed Hamilton Standard props. It was bought by the Marine Corps, which used one to take off from an aircraft carrier; the United States Navy, which fitted it with skis; the United States Coast Guard, to use as Hurricane Hunters; and the United States Army, to recover special forces behind enemy lines. The C-130 appeared in so many other conflicts that they became synonymous with American involvement worldwide. The A and B models were used heavily in Vietnam, not only as troop transporters, but also as "gun ships." With each successive model, the design changed — fuel tanks mounted between engines, more powerful Allisons, more horsepower, higher maximum takeoff weight, increased range, stretched super tankers, "glass" cockpits, and even Rolls Royce engines for the J version. With the C-130, various cargo off-loading techniques became standard: the ground proximity extraction method (a hook attached to the cargo pallet engages a cable on the ground), and the Low Altitude Parachute Extraction System (LAPES). The first civilian Hercules, the L-100, was certified in 1965. Later models had fuselage "plugs" inserted and were used by several cargo carriers, making the fortune of Pacific Western Airlines in Canada.

The first CC-130 to come to the Canadian Forces was the B model in 1960 (since traded back to Lockheed). The thirty-two CAF Hercules, now in the E and H versions, have been used by six squadrons (413, 435, 424, 426, 429, and 436). Besides the tactical and airlift roles, they serve in air search and rescue, resupply of the Canadian Forces station at Alert, and air-to-air refuelling of fighters. The air force freighter has been used in so many disaster relief roles — dropping food supplies in Africa, airlifting refugees in Kosovo's Operation Mikado — that it has taken on a humanitarian image and will be difficult to replace.

The latest addition to the 130 series is the J model; with its six-bladed props, it travels farther, flies faster, and takes off and lands in shorter distances. Fifty years later, the initially unloved ugly duckling remains in production at Marietta, proof that the true test of an aircraft design is timelessness.

De Havilland Buffalo.

DE HAVILLAND BUFFALO

In the early 1960s, the American army's increasing involvement in jungle warfare in both Central America and Vietnam kept De Havilland Canada in Downsview, Ontario busy. The American aviation industry had nothing like the DHC-2 Beaver, DHC-3 Otter, and DHC-4 Caribou. Their short takeoff and landing characteristics, originally developed for the bush pilot market, had earned the Toronto company praise and export orders worldwide, especially from the American military. In 1963, to further develop the Caribou, both the American and Canadian governments agreed to share the cost of redesigning it to be the Caribou 2, later called the Buffalo.

The Caribou was De Havilland's success story — over two hundred had been exported, and it remained in production when the Buffalo was being built. But while its Pratt & Whitney R-2000 Twin Wasps were adequate for its requirements, they dated from the 1940s, and the company wanted to produce an improved air frame with turbo-props instead. General Electric was licence-building British turboprop engines, and its T64-10 was rated at 2,850 hp and was capable of more than 4,000 hp. The new engines were tested in a Caribou airframe through September 1961, while the Buffalo's body was designed. Wider and longer than the Caribou, it sported inverted gull wings that had been moved aft and a completely new large tail. De Havilland's chief test pilot, Bob Fowler, took it on the first flight on May 9, 1964, and pronounced its handling excellent. The aircraft's payload was over 5 tons, its internal volume up to 1,580 cubic feet, and it could carry 44 passengers.

But try as De Havilland did, the Buffalo's full potential would never be fulfilled. The American aviation industry pressed the United States Army to buy locally, and with the development of better helicopters, the American military lost interest, buying only four Buffalos designated as C-8s. The Canadian Forces bought fifteen to be used as transports and search and rescue aircraft, but without the American orders, the aircraft's major market was the developing world — Brazil, Peru, Zambia, Abu Dhabi, Ecuador, Togo, and Honduras. One Buffalo was involved in a major crash when, on April 27, 1993, while carrying its national football team to the World Cup, a Zambian Air Force Buffalo crashed into the ocean off Abidjan, killing all on board. In service with the Canadian Air Force, the Buffalo designated CC-115 was scheduled to be replaced in search and rescue by the CH-149 Cormorant when that helicopter became available. After the 123rd Buffalo was built in 1986, De Havilland ended production.

De Havilland Buffalo.

LOCKHEED CF-104 STARFIGHTER

Timing is everything, especially for aircraft. The F-104 Starfighter, conceived by Lockheed as a futuristic design, was a disappointment to the United States Air Force. It would have been consigned to the breaker's yard and written off, except that its ignominy coincided with NATO's requirement for an "off the shelf" fighter that could carry its nuclear deterrent.

The legendary Clarence "Kelly" Johnson began the Starfighter project as a private venture after talking with fighter pilots in Korea. They wanted a weightless, simple fighter that could outperform the heavier Soviet MiGs. Influenced by the Douglas rocket ship Stiletto, the Lockheed team came up with a similar fuselage that tapered to the nose, flanked by straight, short, ultra-thin wings, their leading edge so sharp that they would need felt covering strips so ground personnel wouldn't cut themselves. It weighed half what the other fighters did, and all of its weight was in the fuselage, since there was no space in the wing for fuel tanks, landing gear, or weapons. Inevitably called the "missile with a man in it," the F-104 Starfighter was the first operational interceptor capable of Mach 2. The inaugural flight took place on March 4, 1954, and the United States Air Force initially ordered 722 aircraft. It did not take long to realize that the pencil-thin fighters had no all-weather proficiency or endurance. The order was cut to 290 aircraft, most of which ended up with the Air National Guard when delivered. Lockheed's manned missile looked destined for an early death, when fate intervened

In 1955, smaller NATO countries like Canada, Italy, Greece, Belgium, the Netherlands, and others that could not afford to create their own tactical nuclear bombers were looking around for an "off the shelf" solution. With the

CF-104 Starfighter.

RCAF, Royal Norwegian Air Force, and Luftwaffe, many also wanted to go supersonic. It was a contract worth millions of dollars; the British offered their Electric Lightning, the French the Dassault Mirage III, and the Swedes the Saab Draken. But none could match the United States for almost giving away the surplus F-104s. Resurrected, the Starfighter became the West's most important fighter; it was bought by all of the NATO air forces and built under licence in Canada, Italy, and Japan. The Starfighters were still flying in the 1990s, the NATO air forces having passed them on to still more countries — Pakistan, Taiwan, and Jordan.

Canada was the second country to select the F-104 to replace its Sabres in Europe. On August 14, 1959, Canadair was given the contract to manufacture two hundred CF-104s and parts for the Luftwaffe's Starfighters. Orenda Engines of Toronto was also licenced to build the J-79 engine. The Canadian Starfighters were not multi-mission aircraft like the American versions but were designed solely to carry nuclear deterrent; their undercarriage and tires were strengthened for this and their cannon sacrificed for more fuel cells. The first Canadian-built CF-104s flew on August 14, 1961, and the last of the two hundred was delivered to the RCAF in 1964. In 1970, the Trudeau government ended Canada's nuclear capability, and within two years, with cannon and rocket launchers installed, all the CF-104s were converted to a ground attack role. As CAF squadrons were disbanded in Europe, the CF-104s were given to Denmark and Norway. Of the 239 used by the RCAF, 110 were lost in accidents — a record not uncommon for all the countries that flew the Starfighter. The Americans dismissed the Starfighter as a "second lieutenant's plane"; the Luftwaffe pilots who crashed the most Starfighters grimly called theirs the "Widowmaker." But the Canadians retained an attachment for Canada's first (and last) nuclear bomber.

Courtesy of the Department of National Defence

In 1983, the CF-104s were replaced by the CF-188 Hornet, and the last Starfighter was phased out of 441 (Silver Fox) Squadron on March 1, 1986. The Turkish air force was given the remaining fifty-two operational CF-104s. In a way, it was fitting that the only aircraft Canada ever had to deliver its nuclear deterrent ended its operational life at almost the same time that the Warsaw Pact did.

CF-104 with Hawker Sea Fury and CF-5.

CP Boeing 737 in the 1968 orange livery.

BOEING 737

Buried in the DNA of the Boeing 737 are the ingredients that made the DC-3 immortal — a classic design that ensures its owners profitability. Until the "Baby Boeing" appeared, the airlines' best customers were invariably senior executives and once-a-year vacationers. The 737 revolutionized commercial air travel, becoming the equivalent of the crosstown bus. In the 1960s, with the decline of the railways, manufacturers built aircraft to cash in on businessmen and salesmen who shuttled daily between cities.

Occupied with its 707, Boeing was a late entry into to the short-haul jet airliner field, and the company only announced that it would build the twin-jet, short-range transport in 1964, by which time its rival Douglas already had its DC-9 Series 10 off the ground. The eighty-five-passenger 737 was designed to complement the long-range 707s and configured to utilize as many components off the 707 assembly line as possible — six-abreast seating and the same equipment in the cargo hold and galleys. Even as Lufthansa, the launch customer, took delivery of the first 737-100 on December 28, 1967, Boeing was already lengthening the fuselage by 6 feet to accommodate 100–124 passengers for United Air Lines — a process that continues forty years later.

The year 1968 was a memorable one for Canadian Pacific Airlines. In keeping with the rest of the Canadian Pacific transportation empire, its beaver trademark and "spaghetti writing" logo vanished, to be supplanted with the eye-catching orange livery and Multimark insignia. It became CP Air, its "Big Orange" aircraft now clearly identifiable anywhere in the world. With its lopsided international route network, the airline embraced the "hub and spoke" concept that would bring passengers to Vancouver and needed short-haul regional jets. The six old DC-6Bs were sold off and, on October 22, 1968, replaced with the first Canadian-registered Boeing 737, "CF-CPB." CP Air and an upstart carrier in Calgary called Pacific Western Airlines (PWA) bought 737s in great batches, in both standard configuration and Combis (half freight), adapting them for urban and Arctic routes. CP Air refined a few to be "executive jets" on April 1, 1970, and put them on seven daily transcontinental flights catering to businessman. PWA dressed their stewardesses in cowboy outfits and created the "Chieftain Airbus" for the Calgary Stampede. For the next thirty years, CP Air (later Canadian Airlines, and finally Air Canada) would standardize its regional fleet on the Boeing 737, the DC-3 of the jet age. The short-haul, low-fare, high-frequency airline phenomenon was pioneered in the 1970s by People's Express in the United States and expanded by Herb Kelleher's Southwest Airlines. In Europe, the Irish Ryanair followed suit, and in Canada, Westjet took on Air Canada's monopoly. All attributed their success to two things: the "no frills" concept and the use of one aircraft — the Boeing 737. When it took control of Canadian Airlines in 1999–2000, Air Canada inherited thirty-two veteran CP Air/PWA/Canadian Airlines 737s. While some were stored in Calgary or scrapped, in 2002, Air Canada used four 737s to launch AC Jetz, a service exclusively for sports teams and corporate business. By 2002, Southwest Airlines operated 355 737s, and Westjet had the 737- 700 model. The latest 737s have improved aerodynamics, flight deck dashboards from the Boeing 777, and can carry up to 189 passengers. But they are still the progeny of the best-selling jet airliner of all time: the classic 737.

Same Boeing 737 in 1998 Canadian Airlines colours.

CF-5 FREEDOM FIGHTER

One of the reasons that the Liberals came to power with an overwhelming majority in 1963 was the confused defence policies of Prime Minister John Diefenbaker. Cancelling the Avro Arrow and buying Bomarc missiles and Voodoos from the Americans had turned many Canadians away from the Conservatives, and the Liberals promised a complete revamping of the military.

Lester Pearson's new minister of defence, Paul Hellyer, sought to do just that, unifying the armed forces and emphasizing peacekeeping rather than nuclear deterrence by using cheap, flexible forces. To replace the Canadair Sabres and CF-100s, the RCAF favoured the F-4 McDonnell Phantom, which could be built with the British and Americans to offset cost. Instead, in July 1965, Hellyer chose the Northrop F-5 Freedom Fighter, reasoning that it was cheaper and less sophisticated than the Phantom and could be built by Canadair and exported to the Third World. Developed in 1959 by Northrop as a light tactical fighter, the F-5 was to attain an almost cult status in the world of affordable fighter aircraft, and it would be flown by American friends and foes alike. If the senior RCAF officers were disappointed by the minister's choice, they were soon occupied with Hellyer's unification program, and, compared with the cost-cutting of equipment in the army and navy, even the lightweight F-5 was better than nothing. The single seat version was designated CF-5A, the dual CF-5D, and 115 planes were ordered. Destined to be forever in search of a role, the CF-5 had its good points. Powered by the Toronto-built Orenda J85 engine then in production for the Tutor, the CF-5 had more speed and a better climb rate than its American counterpart. Because of Canada's NATO commitment to providing air support in Norway (which also had F-5s), all CF-5s were also equipped with in-flight refuelling probes on their star-

CF-5 Freedom Fighter.

Courtesy of the Department of National Defence

CF-5 Freedom Fighter.

board side, additional armour, strengthened windshields to protect against bird strikes, and improved gyro-optical gun sights. Canadian squadrons began receiving their CF-5s, now designated CF-116, in November 1968, and Canadair was given an export order from the Netherlands for seventy-five single seat and thirty dual NF-5s.

The CF-5's deficiencies toward Canadian defence obligations were obvious. It required extensive in-flight refuelling to fly the Atlantic Ocean to Norway and, once there, was too light to carry enough ordnance or fuel to defend itself against the Warsaw Pact, let alone act as an attack aircraft. As well, the Trudeau years were forcing the closure of so many fighter squadrons that when the last CF-5s came off the Canadair line in 1975, they were immediately put into storage.

The remaining CF-5s found roles as trainers for the CF-188s and, in 1988, were expensively overhauled by Bristol Aerospace of Winnipeg to become Hornet simulators. That program was halted in 1995, when the Department of National Defence downgraded its fighter strength by 25 percent, and the CF-5 was completely withdrawn from service to be put into storage and hopefully sold overseas. The export market for the CF-5 had never materialized, primarily because the United States had made available so many F-5s to allies and potential allies in the Third World. When Canadair did secure an order to sell to Venezuela, Northrop took legal action to prevent it. A good light tactical design, the CF-5 had the misfortune to span the country's muddling years as it searched for a role within NORAD and NATO.

CP Air 747-200.

Courtesy of Canadian Airlines.

BOEING 747:
THE FIRST JUMBO JET

On September 30, 1968, one day ahead of schedule, Boeing rolled out the 747. No aircraft this size had ever been seen, and it dwarfed even the specially built thirty-two-ton tractor that towed it out of the plant. Gathered at the Everret assembly plant were the representatives of the twenty-six airlines that had already purchased it, all no doubt craning to spot their company's logo on the fuselage. The era of mass transport by air had begun. Air Canada purchased three 747s and held options for three more; on hand were Herb Seagrim, executive vice-president; Kelly Edmison, regional operations manager; and Ken Rutledge, manager of aircraft contracts. To christen the behemoth were twenty-six stewardesses, one from each purchaser. The women stood on a two-storey structure to reach the "bubble," stepping forward to break bottles of champagne along the nose. The audience listened to American Secretary of Commerce C.R. Smith (the man who had founded American Airlines) deliver the main address.

Later, Seagrim gave an interview to the Canadian press: "Quite apart from sheer size I can't help but be impressed with the airplane's functional qualities all of which remain comfortably within the known state of the art but promise to provide safety, comfort and reliability beyond the fondest dreams of our passengers." The press was warned that the interior of the monster was a far cry from what it would actually be like when the aircraft entered service in 1969, as the entire twenty-foot wide cabin was then crammed with instruments for the flight test program. But that didn't stop everyone from trying to peep in at the star attractions — the spiral staircase that joined the forward cabin to the flight deck and a small passenger area, which the press release said "could be furnished as a private stateroom with bed, a business office or even a theatre or lounge." Some airlines would actually install a bar and piano until seat-mile economics forced them to fill the "bubble" with passengers.

The public wondered how the world was going to accommodate the 747. Everyone had read the comparisons — that the Wright Brother's first flight could have taken place comfortably within the 185-foot cabin, that each Pratt & Whitney JT9D engine weighed 8,430 pounds, cost $800,000, and inhaled 1,600 pounds of air a second (the Conway on the Air Canada DC-8 only gulped 280 pounds a second), that each of the three passenger doors was 42 inches wide, allowing 2 people to board simultaneously, and that it was going to take only 10 minutes to board all 368 passengers.

But it was those 368 passengers (and their luggage) that the 747 was going to disgorge at airports that concerned some people. Travel by air was not meant to happen on such a scale, and airport planners were worried. They even calculated that if each of those passengers on the 747 was sent off by 3 relatives and well-wishers, it would equal 1,104 bodies per flight. At peak hours or during delays, what would be the effect of those numbers on current airports, on parking, fast food outlets, and toilets! It was all in the future. There would be other "jumbo jets," but none as recognizable as the first one — the Boeing 747. The 747 itself would fly for the first time on February 9, 1969, and be delivered to Pan American Airways on January 21, 1970. Air Canada would be the launch customer of the next model, the 747-200M, receiving it on March 7, 1975. CP Air bought four 747-200s from Boeing at $26 million each. The first was delivered to its Vancouver base on November 15, 1973, the employee newsletter exulting that "...the orange giant was their very own ticket to the jumbo era." Unlike Air Canada, CP Air christened and re-christened their 747s in various models several times over. Sometimes they were Empresses of Japan/Italy/Australia/Asia, other times they were named after the great Canadian bush pilots — Russ Baker, Max Ward, and "Wop" May. Some

served in the original orange livery, then in 1987 under the Canadian Airlines Multimark logo, and finally with Air Canada when the airline was bought in 1999–2000. The 747 made flying affordable — it will always be remembered for giving the common man wings.

Courtesy of the Archives of Canadian Airlines

Canadian Airlines Boeing 747-400.

CP-140 AURORA

Lockheed had done it before. In 1938, when its Super Electra failed to live up to expectations as an airliner, it was re-marketed as the Hudson coastal patrol aircraft for the RAF and the RCAF. Twenty years later, when the company entered its Electra turboprop airliner in a United States Navy competition for an ASW platform, the P-3 Orion was an "off the shelf" solution sure to please politicians and pilots alike. Other countries have also used former airliners as ASW platforms, notably Britain's BAe Nimrod and Russia's Ilyushin IL-38. But the best seller, without a doubt, was the Lockheed Orion, and countries with a long coastline to protect — from Japan to Norway, New Zealand to Iran — equipped themselves with it. In 1980, to replace its Canadair Arguses (which had replaced the Lockheed Neptunes) Canada bought P-3s designated CP-140 and called them Auroras, after the Greek goddess of the dawn and new beginnings. The Canadian Forces had wanted a minimum of twenty-four P-3s, but the Trudeau government limited it to eighteen instead. With five based at Comox, British Columbia and thirteen at Greenwood, Nova Scotia, the Auroras maintain a constant patrol of Canada's coast, able to fly over fifty-five hundred miles without refuelling. Equipped with sonobuoys, FLIR (Forward-looking infra red cameras), MAD (magnetic anomaly detectors), ESM (electronic support measures), cameras (both fixed and hand held), plus torpedoes, flares, and the option for air-to-surface missiles, the Aurora carries a minimum crew of ten. By 2002, the Aurora's computers, radar, and instruments were twenty years old, and the patrol aircraft was ready for a complete avionic upgrade. While externally the CP-140 will look the same, the Aurora Incremental Modernization Project (AIMP) will, when it is completed, have overhauled the

C-P 140 Aurora.

Courtesy of the Department of National Defence

Aurora CCX89-266.

computer, radar, and sensor systems throughout the aircraft. The Aurora will no longer be tasked only for maritime patrol missions but will become Canada's long-range patrol and strategic reconnaissance aircraft, able to support joint operations inland, as it did over flood-devastated Manitoba in 1998.

The last three P-3C airframes were taken off the Lockheed production line and outfitted by IMP Aerospace Ltd. of Halifax as arctic and maritime surveillance aircraft. Called Arcturus, they are designated CP-140A.

De Havilland Dash 7.

DE HAVILLAND DASH 7

The 1970s were the years that short takeoff and landing (STOL) aircraft came of age. Until then, STOL aircraft in the public's mind were the Fieseler Storch, the De Havilland Beaver, or the Pilatus Porter, all at home in the mountains or jungles. But for many urban planners, adapting their aerodynamic magic to fly up to fifty commuters at a time in and out of downtown centres became the panacea for urban congestion. STOL air services sprouted from airports like St. Helen's Island, Montreal, Toronto Island, Docklands, London, and Belfast City Centre, and the aircraft invariably seen at all of them was the De Havilland Dash 7, the world's first STOL airliner. Several countries had looked at developing their own commercial STOL aircraft — the Australian GAF Nomad, the German Dornier 128, and the Israeli Arava were the best known. De Havilland Canada had more experience with STOL than any of them, from its Beaver in 1947 to its Twin Otter, then in production. In 1972, the company began a quiet, environmentally friendly STOL airliner project, a high wing monoplane that, unlike the preceding DHC aircraft, was a real passenger aircraft: it had four engines and a retractable undercarriage. De Havilland's Bob Fowler and Mick Saunders test-flew the imaginatively named Dash 7 on March 27, 1975. The Department of Transport certified it for a seven-degree, thirty-foot glide slope and a thirty-five-foot landing reference height, and the first production model was delivered to Rocky Mountain Airways three years later. Like the Beaver and Porter, it achieved STOL by optimizing utmost lift at minimal speed. De Havilland had developed flaps that covered three-quarters of the trailing edge of the wing, extending the rear of the flaps even further with hydraulic jacks. On landing, the hydraulic pressure was jettisoned, and the slipstream retracted the flaps, reducing lift and increasing braking.

Courtesy of the Bombardier Archives

Dash 7 in the Alps.

Other STOL aids included different wing profiles for the outboard and inboard wing, drooped leading edge, and the slipstream of the Hamilton Standard propellers that covered most of the wing, increasing lift and, during the landing flare, reducing power that "killed" lift. Using only 1,200 RPM for takeoff made it a "quiet, environmentally friendly" airliner in urban areas. Residents around STOL ports appreciated that it could rotate at eighty-five knots and land at eighty-three knots — seventy knots if empty. In 1979, the Canadian Armed Forces bought two Dash 7s, designated CC-132, to use in Europe until 1985. Environment Canada had an extended-range Dash 7R built for ice reconnaissance along the Labrador coast. By then, production of the 7 had ceased, as De Havilland Canada was building the faster, more economical Dash 8. But by then, the aircraft that was created for city centre–to–city centre flights could be found anywhere in the world — on the tundra, in Arabian Gulf oil fields, and at classy ski resorts. Air Kenya, Air Nuigini, City Express, Air West, Spantax, Emirates, and Air Tindi were some of the other customers; the Dash 7 fitted in on the Mt. Kilimanjaro route as easily as it did one at Vail or Nuuk (Greenland). British airline Brymon Airways made their fortune flying Dash 7s out of Plymouth Airport to London City Airport.

Perhaps the most appropriate use of the Dash 7's forte was by Tyrolean Airways. Part of Austrian Airlines, the alpine company bought its first Dash 7 in 1980 and its last in 1989. No one who has ever flown into Courchevel Airport on the Swiss/French border will ever forget the experience of the Dash 7 negotiating the "ski jump" used for a runway. The Dash 7 taught the world what STOL was all about.

CT-155 HAWK

British trainer aircraft had been traditionally named after the country's universities (Oxford, Tutor, or Provost) or had some educational attachment. Built completely for the RAF, they always had side-by-side seating, a concept that instructors felt didn't prepare the pilot for solo flight. The Hawk was to be very different. In 1969, the British Ministry of Defence issued a requirement for a tandem seat, single-engined subsonic trainer that had some weapons capability to replace the Folland Gnat then in RAF use. The creators of that aircraft were the British Aircraft Corporation (BAC) and they submitted the Hawk design, in which some of the Gnat's ancestry can be seen. Developed during the earliest period of aeronautical cooperation with the French, which would lead to the SEPECAT Jaguar, the Hawk was the first British aircraft to use metric rather than imperial measures, and rather than the Rolls Royce Viper engine, the Turbomeca Ardour 151 was selected. More expensive, the Ardour was chosen for easy access and serviceability. The aircraft had several improvements over the Gnat: the cockpit canopy was side-hinged, the cockpit was larger than usual and had bigger ejection seats, and the back seat was stepped up to allow the instructor to see over the student's head. Agile, responsive, and sporty, the Hawk could also go transonic in a dive. The first Hawk was delivered to the RAF in November 1976, later replacing the Folland Gnats in the RAF's "Red Arrows" aerobatic team. Export sales were slow at first, as the Alphajet was also on the market at the time. After Finland, Kenya, Indonesia, and Zimbabwe bought Hawks, British Aerospace broke through the Middle East market with sales to Dubai, Saudi Arabia, and Kuwait (where they were temporarily captured by the Iraqis). The United States Navy had a variant, built by McDonnell Douglas as a carrier trainer, known as the Goshawk. In 1987, British

CT-155 Hawk.

CT-155 HAWK

Aerospace (formerly BAC) brought out the Hawk 100, with the more powerful Ardour 871 engine and avionics more suitable for a fighter aircraft than a trainer. It was the Australians who first selected it as lead-in aircraft for their F-18s. So similar were they to front-line fighters that in 1997, the Hawk 115 was chosen for the NATO Flying Training in Canada; the first of twenty-one was delivered to 15 Wing Moose Jaw and 4 Wing Cold Lake in July 2000. Designated CT-155, the Hawks are painted an impressive dark blue, with a red maple leaf overlapped by a NATO four-pointed star on the tail. The students graduate to them from turboprop tandem seat Raytheon T-6A-1 Harvard II trainers used for introductory flight instruction. The Hawks are seen as the most suitable replacement for the aged Tutors of the CAF "Snowbirds."

CC-150 Polaris.

Courtesy of the Department of National Defence

CC-150 POLARIS

Canada's air force was in the forefront of jet transport when in 1951 it bought two De Havilland Comets for military use — long before the British or Americans. It had almost had a troop transport aircraft also named Polaris; the story went that when Canadair was casting around for a name for the DC-4 hybrid in 1946, the choice was between "Polaris" and "North Star." The latter was chosen, and the historic North Star served in the RCAF until 1965, when it was replaced by the Yukon. The five Airbus A-310s were acquired in 1992 from the defunct airline Wardair International by the Canadian Armed Forces for use as long-range transports. Christened "Polaris," all are based at 8 Wing, Trenton, Ontario, in 437 Squadron. Also used for regal, vice regal, and prime ministerial trips, each aircraft is capable of carrying up to 194 personnel or 32,000 kilograms of cargo in support of Canadian forces worldwide. The Airbus A310-300 was built in sections by a consortium of nations: Aerospatiale, Deutsche Airbus (MBB), British Aerospace, CASA of Spain, and Fokker of the Netherlands. A shorter version of Airbus's best-selling A300, it was launched in 1986 with technologically newer wings, horizontal tail surfaces, and engine pylons able to accommodate whatever engines the customer wanted (in the CAF aircraft's case, General Electric CF6–80A1 engines). The Polaris replaced 437 Squadron's venerable Boeing 707s (CC-137), which had provided the Canadian forces with long-range transport for two decades. Besides shuttling between Canadian forces bases at home and in Germany, two of the 707s had been fitted with Beech 1050 aerial refuelling pods and were used as tankers for the CF-5s. When the CC-137s were retired in 1997, this SAAR (Strategic Air-to-Air Refuelling) capability was lost. The need for a CF air-to-air refueller was never more dramatically demonstrated than when the CF-18s returned home on December 21, 2000,

Polaris in Wardair colours.

from Aviano, Italy, following NATO's decision to return the Balkan airspace to civilian air control; they had to be refuelled in-flight over Greenland by a French Air Force KC-135 tanker.

When the German Air Force contracted with Airbus Germany to modify four of its A310s for SAAR, it was an opportunity that allowed Canada, with its twenty-five years of experience in in-flight refuelling, to join in and have two of its own A310s adapted. With the SAAR-modified Polarises, the CF-18s will be able to fly longer without having to rely on foreign air forces for refuelling. On October 2, 2001, a CC-150 based in Rhein-Main in Germany was tasked with providing Canada's contribution to Operation Apollo, the international coalition against terrorism.

The versatile Boeing 707, CAF long-range transport and refueller.

CF-18 HORNET

Canada was the first country that the F-18 was exported to. The aircraft was chosen on April 10, 1980, by the Department of National Defence over the General Dynamics F-16 to replace the CF-101 Voodoo and CF-104 Starfighter. A year later, the Royal Australian Air Force also selected the F-18 over the F-16 for the same reasons: twin-engined safety over long distances and ground attack capability. The F-18's birth stemmed from the cutbacks in military expenditures after the Vietnam war. To replace its F-4 Phantoms, A-7 Corsairs, and A-4 Skyhawks, the United States Air Force and Navy wanted multi-purpose aircraft — the United States Navy looked for a carrier-borne aircraft that could also be a ground attack and air superiority vehicle — and Grumman offered of a version of its F-14 Tomcat. Congress thought that the navy should look at what the air force had chosen and buy the same aircraft. The United States Air Force had just been through a competition for its main fighter aircraft, and the General Dynamics F-16 had won over the Northrop F-17. The latter, with its twin vertical tails (to offset the vortex flows from the leading edge wing extensions) might have lost out to the F-16 but was a good multi-purpose aircraft. However, the navy was tied to its old airplane maker, Grumman, and was reluctant to consider Northrop, as it had no experience with carrier-based aircraft. McDonnell Douglas did, and they convinced Northrop to collaborate with them in a remake of the F-17 — the F/A-18. Reassured on May 2, 1975, the navy opted for the McDonnell Douglas-Northrop marriage, as long as McDonnell Douglas was the prime contractor. Looking like the YF-17, the F/A-18 was adapted for carrier operations with a strengthened airframe and undercarriage and increased fuel capacity. The YF-17's engine was also

CF-18 Hornet.

Courtesy of the Department of National Defence

Courtesy of the Department of National Defence

CF-18.

replaced with the F404 low bypass turbo fan, which has proven relatively uncomplicated to operate. To operate in all weathers and launch radar homing missiles meant a multi-mode radar capability, and in 1977, the Hughes AN/APG-65 radar was selected, its large dish housed in the enlarged nose.

Designated CF-188, the Canadian F/A-18 was the United States Navy version with the carrier landing system replaced by an instrument landing system and a spotlight on the forward fuselage for night identification of other aircraft. At first the name Hornet was not used officially, as translated into French it was "Frelon," then a helicopter in the French Air Force. Unlike previous fighter acquisitions, or as the Australian government would do, the Canadian government did not insist that the aircraft should be made, assembled, or given major offset programs in Canada. The first CF-18s began arriving at 410 Squadron Operational Training Squadron, Cold

Courtesy of the Department of National Defence

CF-18 Hornet.

Courtesy of the Department of National Defence

CF-18 2001 paint scheme for airshow A/C.

Lake, Alberta from the McDonnell Douglas plant in St. Louis on October 27, 1982, and all ninety-eight single-seaters and forty dual (CF-188D) were delivered by September 1988. Seven fighter squadrons were equipped with the aircraft: 416 and 441 at Cold Lake; 425 and 433 at Bagotville, Quebec; and 409, 421, and 439 at Baden-Soellingen, Germany. It was the CF-18s from the last squadrons that were stationed in Qatar during Operation "Desert Storm" in 1990. With the end of the Cold War, all CAF squadrons were returned to Canada, and their number was reduced to four squadrons (416, 441, 425, and 433). Several of the Hornets were also put in storage, in a move to extend the aircraft's use well into the next century.

By 2000, the F-18 had been in service for twenty years and was considered to be in its mid-life. Like the Canadian Armed Forces, the Australian Air Force and the United States Marine Corps and Navy embarked on modernization programs that would keep the fighter effective for another twenty years. Because the Australian aerospace industry had provided much of the electronic components, avionics, and software for their F-18s, the RAAF aircraft were already at the operational standard of

Courtesy of the Department of National Defence

CF-18.

the later C and D models. In Canada, with the squadrons serving in the Balkan conflict, the modernization had to be more extensive. The CF-18 was to have its weapons suite of sensors and air-to-air and air-to-surface missiles upgraded. The key to an avionics upgrade was revamping the 1970s computer system and its software with the ECP 583 (Engineering Change Proposal), which would allow the installation of Jam Resistant Radios, the APG 73 radar, and a Combined Identification Friend or Foe Interrogator/Transponder. The final program for the CF-18 would be the installation of the Air Combat Manoeuvring Instrumentation (ACMI) that would have wing tip mounted pods to record all events such as simulated missile firings, thus allowing the aircrew to recreate the whole mission.

CF-18.

CF-18 (2001 air show demonstration).

CF-100 CANUCK

Canadian born, Canadian bred, and, except for some fifty-three sold to Belgium, used only by the RCAF, the CF-100 was the only indigenous fighter to attain operational use. Known as the "Clunk" because of the sound that the landing gear made as it retracted into the wheel well, it was regarded by Canadians with the affection that one reserves for a member of the family. When A.V. Roe bought Victory Aircraft from the federal government on December 1, 1945, it was a ghost company; its three hundred employees under president Walter Deisher were kept busy with making oil furnaces and truck parts. The federal government wasn't interested in its York airliner, nor was Trans Canada Airlines in its Avro Jetliner, but the RCAF was looking for a jet fighter aircraft. At the same time, the crown corporation Turbo Research Ltd. had built Canada's first jet engine, and the federal government was eager to see it pay its way by using one in a Canadian jet aircraft. In May, Ottawa allowed Avro to buy Turbo Research, and the engines that were being developed were named the Chinook and the Orenda. In August 1946, Avro responded with a design study for a jet fighter, and the RCAF contracted for two prototypes. With the help of British engineers, the XC-100 project took shape, and on December 13, 1949, the prototype CF-100 18101 Mk I was rolled out for taxiing trials. The Orenda engine was still experimental, and Rolls Royce Avons were substituted. Spectators thought the aircraft "teutonic" in appearance, which might have been related to the wind tunnel data used by Avro from captured German documents. The first flight took place on a cold, sunny January 19, 1950, and the CF-100 showed that it could outpace the Gloster Meteor then in RCAF service; it could reach 40,000 feet in two thirds of the time that the Meteor required and immediately set a record of flying from Toronto to Montreal at 638

Avro CF-100 Canuck.

Courtesy of James Floyd

miles per hour. When the Korean War broke out that summer, the government contracted for 124 CF-100s. In February 1951, this was amended to 720, and the aircraft's future was assured.

The Mark 2 flew on June 2, 1951, powered by Orenda engines — the first entirely Canadian-designed and Canadian-built aircraft. The first RCAF squadron to receive them was No. 445 Wolverines. At the coronation of the young Queen Elizabeth II, the CF-100 played its part. Films for televising the coronation on June 2, 1953, were flown by RAF Canberra bomber at record speed from London to Goose Bay, where a CF-100 of 445 Squadron picked them up and took them to St. Hubert. There they were taken by helicopter to the roof of the CBC Radio Canada building in downtown Montreal and televised. The Canuck was the first aircraft to be associated with the great Jan Zurakowski of Avro Arrow fame. He would demonstrate it at the Farnborough Air Show in 1955.

The story went that Walter Deisher had called it the Canuck after the aircraft that his old company, Fleet Aircraft of Fort Erie, built, the Fleet Canuck. The Canuck was the CF-100's official name, but the "Clunk" became the unofficial one. To the hotshot Sabre pilots, she was the "Lead Sled." Whatever the name, until it was retired in December 1962, the CF-100 was the best all-weather fighter that the RCAF had. After it was taken out of front line service, the Canuck served in 414 (Electronic Warfare) Squadron at North Bay, Ontario. When the final "Clunk" was retired in September 1981, it closed forever a chapter in Canadian history.

Challenger with Cosmopolitan.

CL-600 CHALLENGER

Certain aircraft will always be associated with certain prime ministers. Who can forget Neville Chamberlain alighting from the Lockheed Electra, promising "Peace in our time"? Or Mackenzie King consulting his mother by séance before he took a Liberator bomber to England? Prime Minister John Diefenbaker is forever damned for cancelling the Avro Arrow. And from the very beginning, Pierre Trudeau, our first jet-setting prime minister, realized the sex appeal of aircraft and campaigned from a brand new DC-9 against Robert Stanfield's tired old DC-6B. Trudeau's Lockheed Jetstar epitomized glamour, and when Margaret hopped it for New York, there was a hint of scandal attached to it. When Tory Prime Minister Brian Mulroney spent $2.5 million retrofitting his Polaris, Liberal opposition leader Jean Chrétien dubbed it "The Flying Taj Mahal." When he became prime minister, Chrétien would be remembered for another aircraft, the Canadair Challenger. In the 1970s, both De Havilland Canada in Toronto and Canadair in Cartierville, Montreal were perilously close to closing down. With the end of the CF-5 production, Canadair, then owned by General Dynamics, was barely ticking along, dependant for sales on its CL-215 water bomber. The politically damaging layoffs aside, the federal government knew full well that if both companies did shut their doors, the Canadian aviation industry would never recover. As happened with the Avro Arrow, highly trained Canadair and De Havilland personnel would emigrate south. Already both company complexes resembled ghost towns; by 1974, the 9,250 workers that Canadair had employed in 1968 had shrunk to 2,000. With the recession, Canadair's American owners served notice that they were about to pull out. It was even more critical in Toronto, where De Havilland Canada was owned by the British firm of Hawker Siddeley. Selling its own HS 146

The first Challenger was rolled out on May 25, 1978, a propitious day for the future of Canadian aviation.

Courtesy of the Bombardier Archives

airliner, Hawker Siddeley didn't need the Canadian Dash 6 Twin Otter as competition and wanted to put its Canadian subsidiary out of business. To forestall this, on June 26, 1974, the federal government took control of De Havilland Canada, a precedent that wasn't lost on Canadair in Montreal. This, with the growing strength of the Parti Québécois, alarmed the federal Liberals sufficiently to prop Canadair up with $62.5 million in loans on January 5, 1976. But the question was, what should Canadair build to get out of debt? The old standby (i.e. the military market) was longer a sure bet, so in April 1976, it purchased rights to an executive jet developed by William P. Lear Sr. His Learstar 600, with its long-range, high bypass ratio turbofan engines and revolutionary super critical wing showed promise, and, dropping the name Learstar for Challenger CL-600, the company managed to drum up orders for fifty-three aircraft. On October 29, 1976, Jean Chrétien, then the federal minister of industry, trade, and commerce, and Canadair president Fred Kearns announced the start of the Challenger program.

Canadair built three basic versions of the Challenger. The first executive transport (C-GCGR-X) was rolled out on November 8, 1978, a date in Canadian aviation history as momentous as February 23, 1909, when J.A.D. McCurdy lifted off the ice in his Silver Dart. If ever the future of a national aviation industry depended on one aircraft, it was then. The first Challenger would crash in a stall, but between 1978 and 1983, some eighty-three of the basic 600 version were built, until production ceased to begin the 601. The customers had ranged from the Swiss Air Ambulance to the Canadian Armed Forces, which bought two as VIP transports to replace the Convair Cosmopolitans. The life-saving orders came from outside Canada, with Comair (17), Federal Express (25), Lufthansa

Cityline (15), and Skywest (4). The 601, the next generation Challenger, distinguishable by its winglets, was delivered to 412 Squadron on May 3, 1983, and designated CC-144.

Early sales were disappointing with the 601, and by September 1982, Canadair was $370 million in debt and had laid off 1,300 employees. On November 23, the federal government transferred its shares to the Canada Development Investment Corporation and then in 1984 wrote off the debts completely. Freed of the liability and with 148 firm orders, the next year Canadair recorded a profit of $27.6 million and suddenly began to look attractive to other companies. Among these was the Bombardier transportation giant, which purchased it on December 23, 1986, for the bargain basement price of $121 million — a fraction of the company's real value. In August 1987, studies on the next of the dynasty — the fifty-passenger Canadair Regional Jet — were begun. Now in the fifth evolution of the original 600, the 500th Challenger was delivered on May 25, 2000. Twenty years before, no one could know that the Challenger 600 was to be the beginning of the most successful line of executive and regional jets in history. As for Prime Minister Jean Chrétien, on the eve of the Easter weekend 2002, without consultation, he rushed the purchase, through Public Works, of two Challenger 601s for $38.2 million each, with $25 million for pilot training. The government refused to confirm or deny that the two would be equipped with missile detection and evasion features. Predictably, the Canadian Alliance accused him of wanting to fly around in a "Taj Mahal" while the Canadian Forces personnel were making do in the Arabian Gulf with forty-year-old Sea King helicopters. Prime Minister Chrétien's explanation was that the original VIP Challengers operated by 412 Squadron were now almost twenty years old, and one had depressurized over Sweden the previous June when he was on an official visit to Russia. No doubt this is the aircraft that he will forever be associated with.

INDEX